The Educational Technology Anthology Series

Volume One

Interactive Video

EDUCATIONAL TECHNOLOGY PUBLICATIONS
ENGLEWOOD CLIFFS, NEW JERSEY 07632

Library of Congress Cataloging in-Publication Data

Interactive video.

(The Educational technology anthology series; v.1)
Bibliography: p.
Includes index.
1. Interactive video—United States. I. Educational
Technology Publications. II. Series.
LB1028.75.I58   1988   371.3'9445   88-24357
ISBN 0-87778-211-3

Printed in the United States of America.

Library of Congress Catalog Card Number:
88-24357.

International Standard Book Number:
0-87778-211-3.

First Printing: January, 1989.

Table of Contents

All of these articles are reprinted from *Educational Technology Magazine*, published by Educational Technology Publications, Englewood Cliffs, New Jersey 07632. This volume is one of a series of anthologies of articles from *Educational Technology*; for original dates of publication, see pages 191-192.

Part One

General Introduction to Interactive Videodisc Technology and Education

Interactive Video: Educational Tool or Toy?

James J. Bosco

In 1927, John Logie Baird developed a technique for storing television images on phonograph records. Eight years later, Selfridges Department Store in London had prerecorded discs for sale. Since television was in a very primitive state, Baird sold few of his "Phonovision" discs, and the technology for recording television signals on disc lay dormant for about a half century (Schubin, 1980).

During the 1970s, several companies introduced videodisc systems. Disc players such as those sold by RCA use a process reminiscent of Baird's "Phonovision." The RCA players employ a stylus which rides over a grooved and pitted surface. Players sold by Sony and Pioneer use an optical scan process involving a laser. (For a good description of the technical aspects of videodiscs, see Bejar, 1982.)

A videodisc can store about 30 minutes of moving images or 54,000 still frames on one side of a disc. When these images are retrieved with lasers, there is very rapid access to any stored image. The "worst case" access time is a few seconds. Laser discs are extremely durable. The quality of the images is not subject to the deterioration of films, tapes, or slides.

Television commercials have made all of us familiar with videodisc systems for entertainment purposes. This article will consider the use of videodiscs in education.

Most of the interest in the educational use of videodiscs has centered on "interactive video." Interactive video involves the linkage of videodiscs with a microprocessor. This enables the information on the videodisc to be controlled by the microprocessor so that the system can react to learner behaviors. Moving images, still images, computer graphics, and printed information can be combined and structured into an instructional unit which can interact with the learner as he or she proceeds through the instruction. Organizations such as the Architectural Machine Group at MIT, the Nebraska Videodisc Design/Production Group, and WICAT Corporation have been leaders in the use of this technology in instructional applications.

If we could see ten years into the future, we would be able to judge the accuracy of the predictions being made concerning the use of interactive video in education. Will interactive video become a prevalent feature of instructional programs in schools? Or, will it fade away like other highly touted educational innovations? Will it be perceived as a mere toy, something to be played with and then discarded? If interactive video becomes an important feature, will education be the better for its use?

In 1984, there are those who are willing to predict the answers to these questions, and the predictions fall on both sides of the questions. The optimists see interactive video as the wave of the future. The pessimists see interactive video as another panacea that will follow the dubious course of the other supposed innovations such as teaching machines and instructional television. Interactive video, they contend, is just another momentarily seductive trap with no long-term consequence. Such persons would agree with a position expressed by George Hein several years ago (Hein, 1971). Hein argued that the use of "stuff" (his word to characterize the new media) has had little impact because such "stuff" does not cause change with regard to things that matter in education.

What I want to say in this article is grounded on two assumptions: first, that the marriage of the videodisc with the silicon chip has produced an offspring which deserves serious consideration with regard to its instructional use in schools; and, second, that the technology does not contain within itself an assurance of effective use in education. It can just as easily, if not *more* easily, be used in trivial or ineffective ways.

Many of the articles and reports on interactive video which have been produced in the last few years are written from a stance of advocacy. Persons involved with a technology during the early stages of its development are often caught up with the technology as a thing unto itself. Many of the thorny problems associated with the application of the technology in a social context are not considered. Even though these persons produce a literature comprised of ambitious claims, there is value to such discussion, for they may help to stir interest on the part of others more remote from the technology who will ultimately be the persons who make the technology work well or poorly. In order for the technology to be used effectively, we need to get beyond the statements of the first

James J. Bosco is Director of the Center for Educational Research, Western Michigan University, Kalamazoo, Michigan.

generation of advocates to more careful considerations. If interactive video is to become a useful tool in education, and not a mere toy or plaything, we need reasoned analysis as much as enthusiasm.

Current Status of Interactive Video

It is generally acknowledged that early projections for the sale of videodiscs overestimated the market. An article in *Business Week* of March, 1982 quotes the president of GE Video, Inc., Thomas Tucker, who indicated that GE revised its market estimates downward and now believes that no more than six percent of all U.S. households will own a disc player by the end of the decade.

In order to stimulate lagging sales, RCA lowered the price of one of its disc player models to under $300. Disc companies, however, were pleased to find that the disc player owners bought more discs than anticipated (23 rather than the anticipated ten to 12). Most of the discs available on the entertainment market are not interactive, with the exception of a few such as the "Kidisc" and "How to Watch Pro Football."

Even though videodisc merchants have not made the dent in the entertainment market for which they had hoped, they have seen much more action in the industrial market. Major corporations such as Ford, General Motors, and Sears, Roebuck and Company, among others, are using interactive discs for both sales and training purposes. The Army Communicative Technology Office is studying non-print means such as videodisc for presenting technical, instructional, and doctrinal materials.

In primary, secondary, and higher education, experimentation with videodiscs has been relatively limited. The National Science Foundation had been involved in the development of videodiscs, but with the discontinuation of support of science education at NSF, this involvement ended. A new government initiative in science education, the VIM Project in the Department of Education, has disc players in 42 school districts and is planning to build "interactiveness" onto existing non-interactive discs.

One of the disappointments for advocates of videodiscs was the ABC-NEA Schooldisc. This project was intended to produce a number of magazine-type discs with interactive capabilities on a yearly basis. The project was suspended because the cost of producing discs using teams of curriculum and production personnel became too high. Another factor which jeopardized the project was the small number of school districts with disc players. It is sobering to consider that a major television company and the largest educational association in the U.S. have stumbled on a videodisc project.

One of the most comprehensive studies of the feasibility of videodiscs in public education was conducted by the Alaska Innovative Technology Project for the Office of Technology and Telecommunications of the Alaska Department of Education. The reports produced by their study advise caution in moving into the technology (Hiscox and Brzezinski, 1981).

Most interactive videodisc set-ups look impressive. The glamour of the equipment may motivate the student. Thus, even routine instruction may be, at least initially, more interesting to the learner using an interactive disc device because of the machinery. It is perhaps no more improper to use superficial elements of instructional machinery as a motivational force than to design textbooks in an attractive fashion, but interest in the superficial aspects will wear off, and a more substantial basis for use has to be presented.

Education has its technophiles and its technophobes, who mindlessly endorse or resist the use of technology in education. Although they do so in different ways, both camps jeopardize the effective use of the technology. Effective utilization involves an understanding of the points of convergence between capabilities of the interactive video system and the nature of educational tasks. I will discuss each of these and will conclude with a consideration of several issues with regard to the implementation of videodiscs in education.

Replicability and Cost of Instruction

Interactive video enables one to construct an instructional program which can be repeated time and again with total replicability. The use of human sources for transmitting information involves less reliability with each reiteration of instruction. A disc, however, will not modify or delete something it played before because it has become bored, forgotten something, or found a new way to present the information. When it is necessary to present an instructional message to a number of learners in a constant manner, this feature of interactive video is quite useful. For example, when instructing service personnel about a new product, a manufacturer may need assurance that each iteration of the instructional program contains exactly the same information. It is easier to provide an immutable structure on a mechanized than on a human instructional program. The structure provides a sense of assurance that the instructional objectives will be achieved.

When an instructional module will be used repeatedly, the cost of the instruction can be reduced by using an automated system. Even though the development costs for such instruction typically are considerably higher than when using teachers, the costs are successively reduced with each student instructed. This feature has helped to make interactive video attractive to industrial trainers, who must provide the same instruction to large numbers of persons at many sites during the course of a year.

The cost of a disc player, commercial grade, without interactivity, is comparable to the cost of a 16mm projector. The cost for an interactive video system ranges from around $3,000 to $10,000. These units involve microprocessors built directly into the videodisc player or that interface with a microcomputer. The cost of disc players with interactive capabilities is certain to decrease. The price of a 30-minute disc is about $15-$20. The price of a 30-minute 16mm film is at least $150. One difference in the film/disc comparison, however, is that there is something on the order of a $2,000 mastering charge for the first disc. This means that small quantities of discs are considerably more expensive than film, but they become far less expensive in quantity. On a head-to-head comparison, then, videodisc is more economical than film.

Schools have not typically developed their own films but have purchased materials from commercial or educational vendors. For organizations that intend to produce their own programs, the cost of development would be involved. The development of an interactive disc can range from several thousand to several hundred thousand dollars. Production costs for a videotape version of the disc, which is the required first step in the production of a disc, or for a 16mm film, are comparable. The cost of development of a videodisc involves the same considerations as developing a film. The higher the degree of technical quality required, the higher the cost.

When interactivity is involved, two additional costs are incurred. The first of these is the cost of a microprocessor to provide the interactivity, and the second is the programming cost in developing the interactivity on the disc. The cost in programming time for a sophisticated disc can be considerable.

Responsiveness to Human Variability

Concern about the adaptation of instruction to variability among learners is an old problem. In the mid-nineteenth century, rapid urbanization and concomitant growth in public education provoked much discussion and invention centering on the problem of providing mass education in a way which took into account differences among learners. From then until now, a continuing stream of approaches that individualize instruction has been developed.

Videodisc systems provide for individual differences among learners by providing interactivity. Interactivity enables students to adjust the instruction in response to their own needs or capabilities. The predominant mode for adapting to individual differences has been the pacing of instruction. The basis for the design of the instructional sequences has been provided by programmed instruction, which was developed during the 1950s, but the logistics, limitations, and capabilities of programmed instruction when carried by a printed page are considerably different from instances of implementation by an interactive system microcomputer. Work under way at the Architectural Machine Group at MIT on the development of an interactive video training program for repairing an automobile transmission demonstrates how the flexibility of microprocessing can be used to provide instruction that can adapt to the needs of the learner. The learner is not required to push any buttons but rather can control the instruction by touching the screen. A variety of adaptations are possible depending on the needs of the learner. Programs such as this can serve as a model for the use of interactive video in scholastic applications.

The word "interactive" is quite popular in discussions about computer-based instruction. It has assumed a talisman-like quality in the discussions. Because an instructional system is *called* interactive does not mean that it provides a higher level of interactivity than does a well-organized book on the same topic with a good table of contents and index. Indeed, in some interactive video programs, the interactivity appears rather artificially incorporated into the instruction. In some cases, existing videotapes or films have become "interactive" by adding some end-of-section questions with loops for learners who answer the questions incorrectly.

The Active Learner

One of the frequently made points about interactive video is that "... it changes the student from passive observer to active participant" (Anandam and Kelly, 1981, p. 3). In an interactive format, the student is required to make certain motor responses; yet, the equation of motor response with active participation trivializes the notion of what is active and what is passive in instructional situations. Personal experience tells us that there can be a high degree of involvement in a performance even when one is not performing motor responses, such as when attending a compelling play or a concert. By the same token, the provision of some opportunities for a learner to push a button will not save an otherwise boring, uninvolving instructional program. In instances when the need for an interweaving of learner responses with the instruction being presented is germane and not imposed merely because of the capabilities of the system, this feature may well enhance instruction.

The Use of Video Information

There are certain kinds of instructional tasks for which video information is quite useful. For example, an interactive video program on blood circulation developed for elementary students, produced at Simon Fraser University, used animation to show the flow of blood through the heart (Kirchner, 1982). The ability to freeze frames and to use frame-by-frame progressions as well as to repeat sequences in instruction such as this is quite useful.

It should be recognized that unnecessary visual information can be a distraction. Research on the use of comic books in a reading program, for example, indicates that the pictures are detrimental in reading instruction with poor readers (Arlin and Roth, 1978). A verbal description or a still picture of blood passing through the heart, however, would be less expressive of the process of circulation than animated pictures. In such cases, the use of the videodisc is functional rather than merely ornamental.

Assessing and Managing Instruction

One of the features of interactive video which can be useful in some training is the facility to provide a record of the performance of learners. Interactive systems can be programmed to provide output on the progress of the learners through the instructional program. Test scores and a record tracing the learner through the instruction can be provided as a by-product of the instruction. The program can be developed in a way which requires a criterion level of success before the learner can exit from the instruction.

Having considered several characteristics of interactive video systems, let us turn to three elements of the scholastic situation which have relevance to interactive video systems: the chalkboard, the book, and the teacher. The functions that interactive videodiscs may be asked to perform in K-12 applications fall into each of these three categories.

The Video Chalkboard

Instruction is often supported by illustrations, i.e., a map, a diagram, a still picture, a slide, a motion picture, etc. The chalkboard serves here as a metaphor for the teacher's display vehicle. Videodiscs are particularly well suited to serve this function. A one-sided videodisc holds 54,000

images. The price of such a disc should be about $20. The administrator paying for such a disc in 1984 could take comfort in knowing that a teacher in the school district in 2984 could use the *same* disc with no deterioration in quality of the stored images! Single images (maps, pictures) and moving images can be stored on the same disc. Any frame. is retrievable within about three seconds or less.

One can imagine a social studies teacher presenting a class on the USSR. In place of the chalkboard is a large videoscreen. The teacher uses the videodisc at particular points in the instruction to illustrate, amplify, or elaborate on a point. No other system provides the efficiency of storage and the ease of recall as does the videodisc. In work done by the State of Alaska Department of Education, Office of Technology and Telecommunications, the most positive endorsement of videodisc technology was in the replacement of film and slides (Hiscox, 1982).

One shortcoming in the use of videodisc as a comprehensive visual aid system is updating. Currently, the recording of information on a disc could not be done by the teacher (as is possible with videotape). This should be possible, however, in the future.

The interactive feature of videodisc systems comes into play for this application in the ability to access the disc at any point and to move around in any desired sequence. The teacher could program the sequence so that a presentation would be constructed using various segments of the disc augmented with computer graphics or overlays. In a sense, a large, diverse video file cabinet of materials can be made available for the teacher with unmatched ease of selection and use.

The Video Book

Another function a videodisc can serve is to provide options similar to those offered by a reference book. Negroponte (Allen, 1981), for example, has suggested a video encyclopedia. Discussions about the use of videodiscs as reference tools illustrate the gap between abstract conceptualization and tangible realization which often besets the development of a new technology. One of the major criticisms of the video encyclopedia is that although it is technically feasible to transfer the several thousand pages of an encyclopedia onto a disc, the resolution of screens is not as clear as desirable for reading a page. Even though advances in the quality of the resolution of screens will eliminate this problem, it is clear from this comment that the critic is thinking about a videodisc encyclopedia simply as converting to videodisc the printed pages of the volumes. This conception of a video encyclopedia illustrates the way in which thinking about function becomes entangled with form.

Is there really a need to transform information which is in print form in books onto a video monitor? Or, should the technology be available when functions arise which are better served by the video technology than by the book format? Storage and retrieval of information by means of the oral tradition in pre-print cultures provided capabilities and constraints which differ from the printed storage of information. Storage and retrieval of information with the video format provides an altered set of capabilities and constraints. The notion of a reference book using interactive video is an intriguing idea; a clear recognition of what such a "book" would be in actuality is still in a very formative stage, but clearly it should *not* be a "printed-page" book. That is a mis-use of this new, different medium.

The Video Teacher

Consider three teachers. The first teacher lectures to the students and permits no interruptions for questions or comments from students. The second lectures but at particular points in the lecture stops and takes a question from a student, or asks a student to gauge the extent of understanding of the presentation. When necessary, the second teacher will back up and provide additional information. The third teacher is involved in continuous dialogue with students, stopping in mid-sentence to respond to cues that some students are not getting the desired understanding.

Each of these three teachers can be imitated on a videodisc: the first a non-interactive disc; the second on a conventional interactive disc. The third teacher presents more difficulties, but there is nothing to prohibit the design of an automated tutor of this third type—except the needed refinements in programming and instructional design.

An automated tutor can accommodate more than differences in the rate a learner will move through an instructional program. It could accommodate other differences, such as depth of understanding sought by the learner and his or her preferred teaching style. Apart from some notable exceptions, such as the MIT Project, the current generation of discs has not yet begun to contend with these aspects. In some ways, the technology of developing software for automated instruction is parallel to the development of books when monks laboriously copied them. It will not be until the techniques become less cumbersome that we will be able to provide the flexibility and options which the hardware can accommodate.

The Application Problem

Although improvements in hardware such as increased storage time on discs, reductions in the cost of disc players, and the ability to master discs in easier fashion will occur, the essential features of the hardware are in place. Many persons interested in seeing interactive video flourish contend that the need is to solve the software problem. By this they mean that abundant and effective software needs to be available in order to stimulate use of interactive video systems in education.

It seems quite reasonable to argue that the availability of instructional programs which are played on interactive video systems is an important element in the proliferation of such systems in education. Yet, by way of comparison, the availability of many good educational 16mm films has not solved the problem of effective use of films in education. Few advocates of media argue that films, slides, TV, etc., have elevated the quality of instruction in the United States even though there are many good films, etc., available to teachers. Most media experts contend that media have led to less improvement than expected because teachers and administrators do not know how to use the available media.

If interactive video systems are "dropped" into schools in a way similar to films and ETV, ought we not to expect the same consequences? This is much like putting a few word-processing units in a central place and permitting the secretaries to keep their regular typewriters. Such would be likely to produce only modest strides in office automation. Those who have studied the office automation problem have recognized that it is necessary to change work patterns, attitudes, and other features of the human structure of the office in order to capitalize on the efficiency of the new machines.

It is premature to change the structure and practices of schools at this time to accommodate the new technology because there are insufficient discs available. It is not premature, however, to begin to consider how structure and practices would need to be changed in order to improve the quality of instruction using the new technology. The hard problem for superintendents and other educational leaders is not how to get funds to acquire the machinery but how to resolve the personnel and procedural issues raised by the acquisition of machinery.

Schooling generally still means one teacher responsible for group instruction through the use of lecture or discussion, with media, field trips, etc., used as seasoning. If such an approach to schooling is irrefutable or inevitable, then interactive video will be just a new spice jar on the shelf. If, however, we wish to explore the use of automated systems of instruction in ways which substantially enhance quality, then several issues must be faced.

The first of these is the quality and amount of available programs. The instructional programs being developed for training in industry will have little direct impact in education. They will not provide a backlog of discs which can spill over to education. Most of the discs developed for industrial training programs are focused on training needs which have no direct analog in education. The entertainment and mass market for disc players will grow, and some educational discs will be developed. Since there are few interactive disc systems in the mass market, there will be little incentive to develop interactive educational discs for this market. The need is for universities, public school systems, foundations, and government funding sources and corporations with interest in the technology to cooperate in developing prototypical discs. It is reasonable for educators to stand back and adopt a "show-me" attitude. Those persons who believe effective discs can be developed should produce and demonstrate them to cautious but not close-minded educators. The likelihood of the slow development of a market for videodisc systems in education is more of a plus than a minus. It can provide the opportunity to prepare for the technology in a way which would help us eliminate the problems being experienced with the "overnight" emergence of microcomputers in the schools.

Education is a labor-intensive field. Mechanization in industry has been a way of reducing costs. In education, however, mechanization has generally *increased* costs. The mechanisms have simply been added to all other costs—nothing is ever deleted. The justification for purchasing instructional equipment has been that such expenditures will increase the quality of instruction rather than that they will reduce the cost of instruction. It is unlikely that this approach will do much to loosen the purse strings of boards of education now or in the foreseeable future. In order for automated instructional systems to reduce costs, entire courses would need to be delivered using these systems, or modules would need to be produced which could provide a new pattern of faculty deployment.

The development of automated systems of instruction which would replace an entire course of study (e.g., first-year chemistry, fifth-grade math, etc.) is still remote. The development costs for such an ambitious undertaking would be extremely high. In addition, such an approach would require an investment in hardware which is implausible. And even if the resources could be found, there would be considerable resistance from teachers on

philosophic and economic (i.e., job displacement) grounds.

The use of modules is more feasible, but although frequently discussed and used from time to time, the curriculum structure and staffing of most schools present stumbling blocks in using modules delivered by automated systems in other than enrichment or remediation. Since much industrial training is in modules, such as several-hour seminars or training sessions, movement to automated instruction has been more natural in that field.

While the use of interactive video as a "reference book" or as a "chalkboard" raises somewhat less thorny issues of application, the use of interactive video for these functions in education is also problematic. The fact that a new and more efficient means of storing and retrieving information is available will not automatically insure that better information gets stored and retrieved in a way to improve instruction. The real cost of adopting the technology is the cost of the machinery plus the commitment of effort and resources to insure that the new technology will have a catalytic function in causing us to improve the substance as well as the format of instruction. There is just a possibility that the new information technology may bring to educators that question posed some time ago by Norbert Weiner which other segments of our society have already had to face: What can a machine do best; what can a human being do best; and how can the two work in concert?

Many who have written about interactive video speak about the potential of the technology. This expression may be a semantic trap, since it may incline us to think of the technology as having latent power to *become* something. The technology has no latent power. It will be what we make of it. The new information technology has put a new color on the teacher's palette. The quality of the creation which will emerge remains to be seen. □

References and Suggested Readings

Allen, D. Conversation with Nicholas Negroponte. *Videography*, 1981, *6*(10), 44-56.

Anandam, K., and Kelly, D. *GEM. Guided Exposure to Microcomputers: An Interactive Video Program*. Miami, FL: Miami-Dade Community College, 1981. (ERIC Document Reproduction Service No. ED205 238.)

Arlin, M., and Roth, G. Pupils' Use of Time While Reading Comics and Books. *American Educational Research Journal*, 1978, *15*(2), 201-216.

Bejar, I.I. Videodisc in Education: Integrating the Computer and Communication Technologies. *BYTE*, 1982, *7*(6), 78, 80, 82, 84, 88, 90, 92, 94, 96, 100, 102, 104.

DeBloois, M. *et al. Videodisc/Microcomputer Courseware Design.* Englewood Cliffs, NJ: Educational Technology Publications, 1982.

Hein, G.E. The Impact of "Stuff" in the Classroom. In L. Lipsitz (Ed.), *Technology and Education.* Englewood Cliffs, NJ: Educational Technology Publications, 1971.

Hiscox, M.D. A Summary of the Practicality and Potential of Videodisc in Education. *Videodisc/Videotex*, 1982, *2*(2), 99-109.

Hiscox, M.D., and Brzezinski, E.J. *Structure and Feasibility of Group Interactive Disc Systems: Summary Technical Report.* Portland, OR: Assessment and Measurement Program, Northeast Regional Education Laboratory, 1981.

Kirchner, G. Simon Fraser University Videodisc Project: Part One: Design and Production of an Interactive Videodisc for Elementary School Children. *Videodisc/Videotex*, 1982, *2*(4), 275-287.

Schneider, E.W., and Bennion, J.L. *Videodiscs.* Englewood Cliffs, NJ: Educational Technology Publications, 1981.

Schubin, M. An Overview and History of Video Disc Technologies. In E. Siegel (Ed.), *Video Discs: The Technology, the Applications, and the Future.* White Plains, NY: Knowledge Industry Publications, Inc., 1980.

Adopting Interactive Videodisc Technology for Education

Peter Hosie

Introduction

A person making a survey of the literature about interactive videodisc (IV) could easily conclude that a revolution in education and training had taken place. As Bosco (1984, p. 13) observes, "Many of the articles and reports on interactive video which have been produced in the last few years are written from a stance of advocacy." A great number of claims made about IV use in education are speculative. A good example of such exuberance is this comment by Jonassen (1984, p. 2): "There is little doubt that microcomputer-controlled videodisc systems represent the most potentially powerful communication device in the history of instructional communication"; or Young and Schieve (1984, p. 4), "Videodisc technology may well revolutionize education in both public and private institutions by the end of the decade." Such rhetoric is similar to that which accompanied the introduction of microcomputers into schools. Interactive videodisc technology has great potential for education, but there are some important issues to be addressed; progress isn't necessarily assured.

Educators familiar with the plethora of terms used to describe learning involving a computer will not be surprised to find that IV has been categorized in just about all of them; from Computer-Assisted Learning (CAL), to Computer-Managed Learning (CML), to Computer-Based Instruction (CBI), to Computer-Based Learning (CBL), to Computer-Assisted Instruction (CAI), and more. Indeed IV designs and uses have involved elements of all the above. Using Taylor's (1980) tutor, tool, tutee categories, IV best fits the tutor paradigm. But, a point of departure emerges because even the fairly unsophisticated IV's made to date have gone beyond this concept. If a true "marriage" or "fusion" of microcomputers and videodisc has resulted in a hybrid new medium, then an appropriate nomenclature is needed to describe it which is not based solely on computer learning concepts.

Education and Training

While this article deals mainly with the educational potential of IV it is recognized that a grey area exists between "educational" and "training" applications. Many of the issues discussed here affect the development of IV for both education and training.

A number of applications of IV for training have been isolated. IV is particularly effective for teaching mechanical and procedural skills (Priestman, 1984). US military human resource laboratories have found that IV can provide sophisticated simulation training that is more cost-effective than hands-on training in many technical applications (Meyer, 1984). In consequence, the US Department of Defense has taken a leading role in the development of IV in an attempt to find more cost-effective utilization of learning resources. As a result IV has been elected by the US military as its future training delivery medium, resulting in the installation of 50,000 fully interactive systems over five years (*Screen Digest*, February 1986, p. 38). However, the most immediate potential for IV is in sales and marketing. Major corporations such as Ford, General Motors, and Sears are using the system for training and sales.

Communications operators can be taught to operate complex pieces of equipment by IV simulation. Young and Tosti (1984, p. 41) found that "A statistical comparison of learners certified using videodisc simulator equipment showed no difference in the actual ability to operate the complicated communications equipment." Ferrier (1982) described the use of IV for competency-based training; it was judged a cost-effective adjunct for certain applications in training for leadership, management, and organizational development.

Evaluation

Despite the assertions of some authors (e.g., Kearsley and Frost, 1985) there remains a need for credible evidence that such a costly learning system is an effective method of instruction. Much of the rationale for adopting IV is based on intuition.

Hannafin (1985) considers there is a need for empirical research to clearly demonstrate the efficacy of using IV in education. Bosco (1986) claims to present the most comprehensive and recent summary of empirical evaluations of IV application for instruction. Unfortunately, only 16 (8 in schools) of the 28 reports analyzed actually refer to IV uses in education. The remain-

Peter Hosie is Education Officer, Education Department of Western Australia, Kingsley, Australia.

ing 12 reports refer to training applications, mainly for the military. Education deals with a client group that differs in important ways from those receiving training.

Bosco's metanalysis reveals that there is not enough evidence to establish that IV is consistently superior to comparative traditional methods. However, positive results were found for user attitude and training time efficiency. The problem of useful evaluation has more to do with the empirical approach adopted by researchers than the number of studies. Qualitative research strategies have the potential to provide more valuable information for instructional designers and policy-makers.

The *National Interactive Video Case Studies and Directory* (Bayard-White, 1986) is an example of how case studies can provide insight into creative applications of IV. All case studies described start with a detailed analysis of the training needs of the organizations concerned before proceeding with an IV solution. If such information had been available to Bosco to analyze, it is likely that the reason for many of the null items recorded would be apparent. From the information presented by Bosco, it would appear as though the reason for indifferent response by learners to 50 percent of the IV projects considered is that they were inappropriate applications of the technology and lacked engagement for the learner (i.e., poor design).

Doulton (1984) reported on the use of IV in secondary school science lessons in the United States. A comparison of normal classroom experiments with IV simulated experiments was conducted. Results indicated that when IV is integrated effectively, improved standards of laboratory work are evident, as was a greater range of exploration by more talented students. An added bonus is the time saved in comparison with setting up normal experiments; this has important time over task-management implications.

There has been a positive response to the Teddy Bear Disc, which has been produced by the Open University for use at residential summer schools. IV's are seen as a way of augmenting or even replacing summer school laboratory sessions as well as allowing students to participate in experiments otherwise unavailable to them (Williams, 1984).

Teh and Perry (1984) reported the results of an Australian developmental project which designed and evaluated IV-based materials for teaching weather forecasting. IV materials were found to be an effective teaching medium in geography, with trial subjects (who were trainee teachers) achieving superior scores for content understanding.

The Alaskan Department of Education has conducted what is probably the most comprehensive study of the feasibility of using videodiscs in public education. Reports produced by the Alaskan Innovative Technology Project advised caution in moving into this technology (Hiscox *et al.*, 1981). However, positive endorsement was given to the use of this IV to replace film and slides. Other developments, which are a by-product of cut-backs in funding for education, are affecting the adoption of this technology. The American ABC-NEA Schooldisc project planned to produce a number of magazine-style videodiscs on an annual basis. However, when production costs involving curriculum personnel became too high, the project was suspended. A National Science Foundation videodisc development suffered a similar fate (Bosco, 1984). Most educators seem to be adopting a "watching brief" because of the well-founded fear of over-extending limited budgets.

The limited data available suggest that IV is an effective teaching-learning medium, especially in the area of training simulation; removing the variability of human teaching is a major advantage of IV courseware. But there is an urgent need for more comprehensive evaluations, especially related to school environments. Learner acceptability of the medium needs to be established. Comparison of IV and traditional teaching methods must provide a detailed analysis of cost-effectiveness.

The Key: Interactivity

Bork (1984) believes computers are going to be an important factor in all human learning because they make learning truly interactive for large numbers of learners on a cost-efficient basis. Bork (1984, p. 3) also observes that "... many of the videodisc plus computer modules produced so far by video people are extremely weak with regard to interaction." He contends that students and teachers are content with very weak forms of interaction because these are such an improvement over non-interactive learning media. Bork's observations seem well founded.

How IV technology can affect learner interaction requires examination. The University of Nebraska Group has defined four levels of interactivity for a videodisc (Hart, 1984; Priestman, 1984). The sophistication of these levels affect the kind of learning possible. Categories which are learner-based, rather than technology-based, are being developed and should provide a more appropriate system of classification.

Design Considerations

Balance and control over learning strategy and content by the student is important. Videodiscs provide a vast information base which is quite different from that available with a computer. The more information available, the greater the flexibility in combining sequences. Evidence suggests (Hartley, 1981) that student control over learner strategy is the most efficient approach to CAL design. Encouraging individual routes through information will assist students to become more actively involved in the learning process.

An unrealized potential exists for learners to control instructional presentations, without lessening the overall coherence of the courseware. Hedberg and Perry (1985) claim that IV has eliminated the requirement for materials to be structured for the learner. They assert that not only has IV improved interactivity with visual materials, which can be incorporated into an instructional sequence, but interactivity beyond the designers' original intention is also possible. While the flexibility of IV design is important, the crucial problem lies in getting teachers to use such software effectively. Open-ended designs are no value under the control of closed minded teachers! Hedberg and Perry (1984, p. 6) also claim that the addition of "dynamic and static visual display enabled students to 'see' events that were not previously possible as part of CAI lessons."

Nievergelt (1982) agrees that programs involving human/machine dialogues should avoid designs which are passive in format. Instead learners should be given as much control as possible over the programs, or at least opportunities for regaining control at some stage of the instructional sequence. While the potential for learner control of well designed IV's is acknowledged and considered desirable, there is insufficient evidence to date to refute or deny these assertions.

Few educationists should need to be reminded that learning is not a passive process. Understanding and knowledge involve active processing rather than passive reception. A consistent claim made in favor of using IV is that "it changes the student from passive observer to active participant" (Anandam and Kelly, 1981, p. 3). The ineffectiveness of modern media as a learning device (in comparison with the written text) is due to the lack of opportunity for interrogation, and allied to this observation is the criticism that the learner loses control over the pace of instruction (Clark, 1984). New theories of learning which will inform instructional design will have to be formulated to allow for these new ways of interacting with the subject matter. Such theories will need to be incorporated into IV designs. Hedberg (1985/86) has suggested a set of design heuristics for IV which encourages student involvement with the process of learning. If the claims made for IV are correct, then learners will benefit by a medium more amenable to individual learning styles.

Microcomputers can provide a more individualized learning experience. CAL learning has largely been developed on the basis of learner interaction. Perhaps the greatest limitation of computer learning has been its delivery—predominantly in text or in diagrams, without the visual and aural attractiveness of television. Film and video do not always lend themselves to informing about higher order concepts, and it is often difficult for a learner to conceive of actions presented in text or diagrammatic form. Educational television has not produced improvements in learning to the extent originally predicted. Perhaps this is largely due to the receptive mode of learning it encourages (Laurillard, 1982). Certainly the ongoing linear nature of educational television does not adjust for the pace and learning style of individual students (Teh and Perry, 1984). Capitalizing on the strengths of the two media, while limiting the disadvantages, should result in a more active learner interaction.

Developing the Innovation

How will IV translate into effective instructional applications? Cost considerations are, understandably, uppermost in most educators' minds. Schools are just recovering from spending a considerable portion of available funds on microcomputers. Besides an inability to afford the hardware, there is a pressing need to retrain teachers to accept and use new technology. Teacher acceptance will be an important hurdle to overcome. Presumably the widespread adoption of microcomputers in schools will ease the initial reluctance, but this may not prevent teachers from feeling that their independence is under attack. Developing highly interactive IV's is a costly exercise. Educational administrators will be keen not to exacerbate the problem by lack of consideration of design quality. It is widely accepted that the majority of educational computer software is poorly designed; if IV is left to be developed by free-market forces, a repeat of this situation is probable.

Arguments which center on increasing the quality of education as a justification for investing in this or any other technology will fall on unsympathetic ears. A number of IV advocates seem convinced that this technology has the potential to not only transform the delivery system of public education, but also to reduce costs (Price and March, 1983). There is potential for reducing the cost of delivering education using IV for well targeted applications. However, if IV is going to be adopted, it will be at the expense of some other aspect of education because proposals which require an increase in the overall education budget allocation are doomed.

The cost of developing IV with even a reasonable amount of interactivity is great in terms of man hours and technology. Media production, especially broadcast quality television, requires a large investment in equipment, and high labor costs. Competent planning is vital with this medium if costs are to be contained and a long shelf life of the courseware is to be assured.

Butler (1981) estimates it would cost between $500,000 (U.S.) and $700,000 to develop enough courseware for a single one-year college course. As Fletcher (1985) notes, instructional videodiscs are produced in small quantities resulting in high unit costs; for example, to produce five or fewer videodiscs costs $615 per side. The unit cost falls to $17.00 per side for quantities of more than 500. These costs are falling at various rates but there is still the large courseware design cost to consider. The cost of courseware design and production will easily overshadow the cost of hardware as a major impediment to the utilization of IV by education.

IV will not become a reality in schools and institutions of higher education unless economies of scale can be developed. How this is likely to occur is not clear. If the lack of mutual cooperation in the production of curricula and audiovisual resources to date is any indication, there is no cause for optimism. The development of a national curriculum for certain subjects would allow funds to be pooled to produce IV's, although local curriculum needs must be considered. Perhaps certain attributes of the technology may be exploited to overcome this problem. A compromise solution to the inevitable conflict of local curriculum idiosyncracies and the need for economy of scale is needed. "Generic discs" (Jonassen, 1984) or a "video databank" (Cohen, 1984), containing material in a variety of forms related to subject areas, could be developed to form a visual database suitable for use in a variety of circumstances. Generic discs could provide video programming at reasonable cost for a wide audience, while allowing for local learning requirements.

A simply formulated authoring system is needed to enable flexible access by the microcomputer to videodisc material. Flexible authoring systems which are simple to use, such as MUMEDALA (Barker, 1984) and MICROTEXT (Barker and Singh, 1984) are assisting in the design of IV's. Hardware specific authoring systems such as PHILVAS (Barker and Singh, 1985) are simple and effective to use but are also expensive and not transferable between different systems. Unfortunately, there is an inverse relationship between the ease of authoring and flexibility of the program.

Authoring languages which provide a very high level of support in the way of prompts and instructions tend to produce less interactive presentations than those requiring programming. Conversely, writing interactive programs in computer languages takes considerably longer, and as a consequence, costs more. The next generation authoring software as detailed by Kozma (1986) will overcome many of the hitherto limitations of authoring systems.

Teacher Acceptability

In the past, innovators in educational technology have assumed teachers would alter their methods to accommodate technological advances. But the lack of software may not be as much of a problem as overcoming teacher reluctance, and retraining. Simply making 16mm films widely available has arguably not resulted in significant learning improvements. Bosco (1984) has suggested that re-organizing certain subjects into modules (similar to industrial training modules) may be a more efficient way of utilizing technologies like IV. Teachers will need to adopt a more managerial, as opposed to an instructor role, if IV is going to be used extensively. IV designs could assist teachers in managing student learning environments; test scores and record keeping are readily organized. Designs could be developed that are suited to criterion-based learning.

The great potential of IV to provide a complete, individualized learning system will only be realized if professional organizations and expertise are rallied to create high-quality software. Some countries have already taken positive steps in this direction. Simcoe (1983) found that approximately 40 percent of American media centers are utilizing some form of IV, with over 25 percent using some level of IV for instructional purposes. The French Education Ministry has already begun supplying videodisc players to schools for use in interactive applications with microcomputers. The National Interactive Video Centre in Britain is stimulating development. A register of research and development, technical briefing, and support for production consortia are given by the Centre. Four Australian university institutions have produced IV's. The Australian Caption Centre has produced the first PAL IV for schools the "Supertext Superdisc . . . Ask the Workers," in cooperation with the Education Department of Western Australia. The "Supertext Superdisc" has been evaluated in detail and will go some way to providing answers to questions posed in this paper.

Conclusion

Combining the visual stimulation of moving and

still images with the interactive capabilities of computer technology has resulted in a potentially powerful learning medium—IV. Before IV can become widely accepted by education, some important issues must be resolved. A substantial amount of high-quality courseware needs to be developed. For the cost of developing courseware to be justified, it needs to be relevant to the needs of a large and diverse student population.

Poorly considered usage of IV will result in inappropriate adoption of the technology. There is an urgent need for effective evaluation to establish the type learning best suited to the medium, and whether cost justification for large-scale adoption can be established. Large-scale adoption of IV will require a substantial commitment of funds, and energy, if software of a high enough quality is to be developed. Teacher resistance and the need for re-training are factors that should not be overlooked. Educational budgeting is in fiscal demise, which means that for IV to be developed, other areas of endeavor may suffer. Decisions of this degree of importance, then, should be based on solid cost-effectiveness analyses and educational rationales. □

References

Anandam, K., and Kelly, D. *GEM. Guided Exposure to Microcomputers: An Interactive Video Program.* Miami, FL: Miami-Dade Community College, 1981, p. 3.

Baker, F.B. *Computer-Managed Instruction: Theory and Practice.* Englewood Cliffs, NJ: Educational Technology Publications, 1978.

Barker, P.G. MUMEDALA: An Approach to Multi-Media Authoring. *British Journal of Educational Technology*, 1984, *1*(15), 4-13.

Barker, P.G., and Singh, R. A Practical Introduction to Authoring for Computer-Assisted Instruction. Part 3: MICROTEXT. *British Journal of Educational Technology*, 1984, *2*(15), 82-106.

Barker, P., and Singh, R. A Practical Introduction to Authoring for Computer-Associated Instruction. Part 5: PHILVAS: *British Journal of Educational Technology*, 1985, *3*(16), 218-236.

Bork, A. Computers and the Future: Education. *Computer Education*, 1984, *8*(1), 1-4.

Bosco, J.J. Interactive Video: Educational Tool or Toy? *Educational Technology*, April 1984, *24*(4), 13-18.

Bosco, J.J. An Analysis of Evaluations of Interactive Video. *Educational Technology*, May 1986, *26*(5), 7-18.

Butler, D. Five Caveats for Videodisc in Training. *Instructional Innovator*, 1981, *26*(2), 16-18.

Bayard-White, C. *Interactive Video Case Studies and Directory.* London: Council for Educational Technology, December 1985.

Clark, D.R. The Role of the Videodisc in Education and Training. *Media in Education and Development*, December 1984, 190-192.

Cohen, V.B. Interactive Features in the Design of Videodisc Materials. *Educational Technology*, 1984, *24*(1), 16-20.

Copeland, P. An Interactive Video System for Education and Training. *British Journal of Educational Technology*, 1983, *14*(1), 59-65.

Copeland, P. Cavis: From Concept to System. *Media in Education and Development*, June 1983, 74-78.

Davidove, E.A. Design and Production of Interactive Videodisc Programming. *Educational Technology*, June 1986, *26*(6), 39-41.

DeBloois, M.L. Principles for Designing Interactive Videodisc Instructional Materials. In M.L. DeBloois (Ed.), *Videodisc/Microcomputer Courseware Design.* Englewood Cliffs, NJ: Educational Technology Publications, 1982.

Doulton, A. Interactive Video in Training. *Media in Education and Development*, December 1984, 205-206.

Dunbar, R. Computer Videodisc Educational Systems. *Australian Journal of Educational Technology*, 1985, *1*(1), 21-38.

Ebner, D.G. *et al.* Current Issues in Interactive Videodisc and Computer-Based Instruction. *Instructional Innovator*, 1984, *29*(3), 24-29.

Ferrier, S.W. Computer-Aided Interactive Video Instruction: Closing the Gap Between Needs and Outcomes in Competency-Based Leadership, Management, and Organizational Development Training. *Programmed Learning and Educational Technology*, 1982, *19*, 311-316.

Fletcher, S. Interactive Videodisc Technology: An Introduction. *Educational Technology Newsletter*, May 1985, p. 2.

Fletcher, S. *Screen Digest*, March 1985, p. 57.

Fletcher, S. *Screen Digest*, February 1986, p. 38.

Gienke, M.A. Cambridge Video Disc Project. Paper for the Australian Society for Educational Technology Conference, September 1984.

Griffith, M. Planning for Interactive Videodisc. *Media in Education and Development*, December 1984, 196-200.

Hannafin, M.J. Empirical Issues in the Study of Computer-Assisted Interactive Video. *Educational Communication and Technology: A Journal of Theory, Research and Development*, Winter 1985, *33*(4), 235-247.

Hart, A. Interactive Video. *Media in Education and Development*, December 1984, 207-208.

Hartley, J.R. Learner Initiatives in Computer-Assisted Learning. In U. Howe (Ed.). *Microcomputers in Secondary Education.* London: Kegan Paul, 1981.

Hedberg, J.G. Designing Interactive Videodisc Learning Materials. *Australian Journal of Educational Technology*, 1985/86, *1*(2), 24-31.

Hedberg, J.G., and Perry, N.R. Design of Interactive Video Materials: Problems and Prospects. Paper presented to the Computer-Aided Learning in Tertiary Education Conference, Brisbane, September 1984.

Hedberg, J.G., and Perry, N.R. Human Computer Interaction and CAI: A Review and Research Prospectus. *Australian Journal of Educational Technology*, 1985, *1*(1), 12-20.

Hedberg, J.G., and Perry, N.R. Learning Task Requirements and the Design of Interactive Video. Paper presented to the Annual Conference of the Association for Educational Communications and Technology, Anaheim, California, January 1985.

Hiscox, M.D. *et al.* Assessment Report 1-7. An Update on the Alaskan Assessment Program. Alaskan State Department of Education, 1983.

Jonassen, D.H. The Generic Disc: Realizing the Potential of Adaptive, Interactive Videodiscs. *Educational Technology*, January 1984, *24*(1), 21-26.

Kearsley, G.P., and Frost, J. Design Factors for Successful Videodisc-Based Instruction. *Educational Technology*, March 1985, *25*(3), 7-13.

Kozma, R.B. Present and Future Computer Courseware Authoring Systems. *Educational Technology*, June 1986, *26*(6), 39-41.

Laurillard, D.M. Interactive Video and the Control of Learning. *Educational Technology*, June 1984, *24*(6), 7-15.

Laurillard, D.M. The Potential of Interactive Video. *Journal of Educational Television*, 1982, *8*(3), 173-180.

Lloyd, P. (ed.). Putting Interactivity into Perspective. *Audio-Visual*, February 1984, p. 27.

Meyer, R. Borrow This New Military Technology, and Help Win the War for Kids' Minds. *American School Board Journal*, June 1984, 23-28.

Nievergelt, J. The Computer-Driven Screen: An Emerging Mass Communications Two-Way Medium. *Educational Media International*, 1982, *1*, p. 7.

Parsloe, E. Learning by Doing. *Media in Education and Development*, December 1984, 201-204.

Parsloe, E. Interactive Video. *Media in Education and Development*, June 1983, 83-86.

Price, B.J., and Marsh, G.E. Interactive Video Instruction and the Dreaded Change in Education. *Technological Horizons in Education*, May 1983, 112-117.

Priestman, T. Interactive Video and Its Applications. *Media in Education and Development*, December 1984, 182-186.

Simcoe, D.D. Interactive Video Today. *Instructional Innovator*, November 1983, 12-13.

Taylor, R. (ed.) *The Computer in the School: Tutor, Tool Tutee*. New York: Teachers College Press, 1980.

Teh, G.P., and Perry, N.R. The Use of the Interactive Videodisc in Teaching Geographic Concepts. Paper presented at the Annual Conference of the Australian Association for Research in Education, Perth, November 1984.

Thomas, W. Interactive Video. *Instructional Innovator*, February 1981, *26*(2), 19-20.

Williams, K. Interactive Videodisc at the Open University. *Media in Education and Development*, December 1984, 193-195.

Young, J.I., and Tosti, D.T. *The Effectiveness of Interactive Videodisc in Training*. U.S. Army Technical Report, 1980.

Young, J.I., and Schlieve, P.L. Videodisc Simulation: Training for the Future. *Educational Technology*, April 1984, *24*(4), 41-42.

Part Two

Interactive Technology

Interactive Lesson Designs: A Taxonomy

David H. Jonassen

Introduction

Perhaps the most lasting effect of computer-based instruction on instructional design and development will be the impetus it has provided to *interactive instruction*. The intense focusing on alternative instructional designs based on the informational handling capabilities of the computer has resulted in a profusion of interactive approaches. As computers, especially microcomputers, become even faster and more powerful, the variety of possible designs should continue to expand.

The current growth of microcomputer controlled, interactive videodisc-based instruction has provided new dimensions in adaptive, interactive designs for instruction. The most significant drawback of this trend to instructional developers, however, is that the resultant instructional designs once again are hardware technology-driven. In the short history of instructional technology, our designs and techniques have too often constituted only reactions to emerging technologies. Technologies have come forth, and we have sought to apply these for instructional purposes. The latest technological bandwagon is interactive videodisc.

The phenomenon of hardware technology-driven design contradicts virtually every premise of the systematic design processes the readers of this magazine expostulate. These principles suggest that we consider the technology (hard technology, delivery systems, and soft technology strategies) only after we have completed front-end analysis. That is, until we refine our instructional needs and goals, we don't consider the sequence, strategy, or delivery mechanism for meeting those needs/goals.

Interactive computer technologies, including interactive, computer-controlled videodisc, have been recognized as so potentially powerful and therefore so compelling that they have become *media in search of designs*. Just what can we do with interactive video?

David H. Jonassen is Associate Professor, School of Education, University of North Carolina at Greensboro.

Ultimately, it is the soft "design" technologies that make the greater difference in learning outcomes. So, in order to re-exert control over hard technologies, we need to think first about interactive designs in a general way. Is an individualized, interactive design required in each particular design situation? If so, we must decide the interactive design that is most appropriate, only then followed by consideration of the hard technology most capable of delivering that interactive lesson. So, rather than creating problems to which we apply our most popular interactive technology, we need to develop design processes which identify the required components of interactive, adaptive instruction.

In order to facilitate such a procedure, what I am proposing in this article is a simple taxonomy of interactive lesson designs which may facilitate the instructional design and media selection processes.

Interactive Instruction Defined

Interactive lessons, contrary to any inferences you may have drawn from the previous discussion, did not originate with computer-based instruction (for example, see Langdon, 1973). Refraining from any attempt to establish the historical or philosophical antecedents of interactive instructional procedures (rooted, inevitably, I suppose, in the practices of a former Greek civilization), let us begin at the start of the age of "instructional technology." Interactive lessons are those in which the learner actively or overtly responds to information presented by the technology, which in turn adapts to the learner, a process more commonly referred to as *feedback*. The point is that interactive lessons require at least the appearance of two-way communication.

Programmed instruction was the first and probably the most significant technology to be identified as interactive. It is interactive because the learner overtly responds to material presented to him/her, rather than passively receiving information transmitted through the medium (e.g., slide, film, television). In early, linear forms of programmed instruction, the mental behavior elicited by the medium was largely recall from short-term memory. Students indicated by their responses that they retained information, at least for the moment, that they just decoded from the text. The mental behavior elicited by such linear materials consisted largely of iterative rehearsal of the information just presented. In an effort to accommodate to the ability levels of different learners, *branching* PI presented *adaptive* interaction, which provided remedial sequences for learners who, by their responses, indicated that they did not comprehend the material. The levels of processing elicited by some branching programs was deeper than the rote rehearsal of linear forms of programming. That is, designers of branching PI sought to assess how well learners actually *understood* what they were reading or seeing.

The most popular theoretical context for considering most interactive learning materials, such as programmed instruction, is the "mathemagenic hypothesis." Mathemagenic (a term deriving from, who else?, the Greeks) behaviors are those which "give birth to learning." The goal of mathemagenic activities is to make that birthing process predictable by identifying the combination of information acquisition processes that will lead most efficiently to learning specific information (Rothkopf, 1970). The major mathemagenic device for doing that is also the primary interactive design tool—the adjunct question. In this research methodology, questions intended to stimulate specific types of information processing are inserted before, after, or within textual materials (Rothkopf, 1970) or audiovisual materials (Dayton, 1977). They are designed to engage the learner in processing the information read or seen in ways that will lead to specific kinds of knowledge or skill acquisition. That is, the learners are required, in order to answer the questions, to think about information in certain ways. Operationally, then, interactive learning designs have been premised on instructional design principles of *task analysis*. Identify and classify the task; then provide relevant practice in learning materials.

Limits of Interactive Designs

The primary criticisms of this approach is the restrictive, behavioral orientation, which is inconsistent with theories of learning and knowledge acquisition (Jonassen, 1985). Learners necessarily must process information actively in order to comprehend and remember it. This process is necessarily individualistic, since learners must access prior knowledge constructs in order to apply them to the information they are attempting to learn. Prior knowledge is based upon the unique set of interactions with his/her environment that each learner has had. So, rather than allowing the learner to generate meaning for information read or seen, based upon existing knowledge, mathemagenic behaviors force a specific type of processing at which the learner may or may not be adept. Too often the mental behavior elicited by mathemagenic activities is just as shallow as the rehearsal implied by linear PI. What is certain, regardless of your theoretical orientation, is that learning is an active process. Therefore, the more mentally active learners are as they process information from learning materials, that is, interact with the materials they are trying to comprehend, the more likely they are to comprehend them.

Adaptive Instruction Defined

The effectiveness of an interactive design is a function not only of the type of student-technology response, but also of the action of the technology as a result of that interaction. So, in addition to the interaction of the learner with the instructional program, the promise of interactive learning programs is the promise of adaptation to the learner. Even Pressey's original 1926 teaching machine offered a low level of adaptation to the learner's ability. That is, when a learner answers an inserted question or performs an inserted activity designed to require specific information processing, the instructional technology should be "smart enough" to evaluate the learner's performance and alter the course of instruction accordingly. We can design interactive learning materials capable of adapting to learner ability, motivation, background, and so on. The ideal Socratic approach to tutorial instruction involves not only the posing of questions and problems for the learner, but also the ability of the tutor to select additional, more relevant questions or to explain, prompt, cajole, or do whatever is necessary to assist the tutee in acquiring the knowledge. Interactive, tutorial instruction is an inherently individualistic process that involves accommodating to individual learners' needs. The point is that when we speak of interactive instructional designs, we are usually referring to interactive *adaptive* designs.

Interactive, adaptive designs are becoming more prevalent, in part because of the adoption of the mathemagenic design model and in part because the emerging computer-based technologies lend themselves to interactive and adaptive forms of instruction. However, as I indicated above, it is heretical to principles of instructional design to begin with assumptions about which medium to use in designing instruction. If, as a result of front-end analysis, we decide that an interactive, adaptive design is needed, we need to look at the instructional requirements in terms of the task/content requirements of the objectives, the type/level of learner-medium interaction, the type/level of adaptation by the medium that is required, and the characteristics of the medium needed to provide that adaptation. What is needed, then, is a planning heuristic to aid in the identification of relevant types and levels of interaction and adaptation. Therefore, what I shall present in the remainder of this article is a taxonomy of interactive, adaptive instructional designs. It does not prescribe any specific medium. Rather, it describes interactivity and adaptation

Figure 1

Taxonomy of Interactive,
Adaptive Lesson Designs

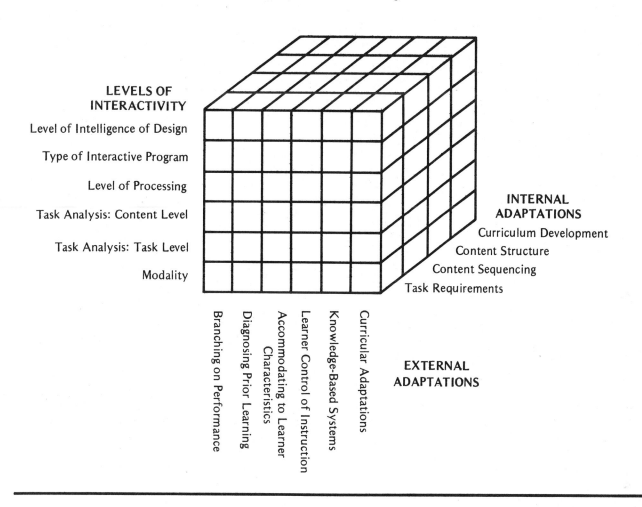

outside the context of any particular medium. The taxonomy is represented in Figure 1 as a cube, the axes of which represent (1) levels of interactivity, (2) levels of internal adaptation, and (3) levels of external adaptation. Each dimension is hierarchical, that is, it represents levels of generality. Each dimension may interact with the other two to describe an interactive, adaptive lesson design.

After constructing and analyzing your instructional objectives and determining that an interactive lesson is necessary, you should determine the types and levels of interaction necessary to support your instruction. Next, determine the level and characteristics on which instruction should be adapted. This includes internal program adaptation to the characteristics of the content as well as adaptation to external characteristics of instruction, *viz.*, relevant learner characteristics. Finally,

these requirements must be filtered through the characteristics of each technology to determine whether or not the lesson design can in fact be supported by the specific technology as it exists today.

Levels of Interactivity

In Figure 1, the vertical axis presents a continuum of interactivity, arranged from specific to general, from the nature of the learner's actions to the orientation and organization of the instructional technology. The most specific interactive characteristics are the modality of the learner's performance while interacting with a tecnology and a task analysis of the performance required by the interaction. The next level of interaction considers the instructional context (type of program) and finally the level of intelligence of the system, that is, the degree to which the technology can internally represent the structure of the knowledge it is conveying and its ability to learn from the learner.

I.1 Modality of Interaction

The modality of the interaction refers to the sensory systems involved in the reception of the message from the technology and the response made by the learner. This modality classification should consider both the sensory modality through which instructional messages are transmitted and the type of message within the modality, that is, digital or iconic messages, since digital messages are processed and stored by an entirely different cognitive system than iconic messages.

While most information is received through visual and then auditory modalities, many interactive media require responses through other sensory systems. For instance, while computer-based instruction (CBI) presents most information visually, it requires tactile responses (i.e., the keyboard), which may facilitate or impede learning, depending upon the nature of the task. Many computer-based instructional systems have begun using touch-screens, and speech synthesis and voice recognition chips are beginning to appear in microcomputer-based systems, so that the instructional modality can be better fit to the task requirements. Similar problems occur with regard to the nature of the symbols employed in learning sequences. I have previewed many CBI programs that employ sophisticated graphics (iconics) to teach visual concepts and then present directions to students in language (digital symbols) several grade levels above the target audience. Even when modality or symbol shifts don't represent an impediment, instructional design principles recommend that the modality and symbols of the response be consistent with those required by the objective, and that the modality of the message be appropriate to the nature of the information being transmitted.

I.2 Task Analysis

The next most abstract way of analyzing the nature of the learner interaction in any lesson is to conduct a task analysis of the learner's behavior. Beyond simply decoding messages through various sensory systems, the learner is assigning meaning to and processing the information in some rote or meaningful way. This is perhaps the most important type of interaction analysis, as it frequently is considered the core of the instructional design process and, at the very least, a major aspect of it (Kearsley, 1978). Task analysis ought to consider both the nature of the learning task and the content level of the material being processed.

I.2.1 Task Level

Learning tasks represent either remember or use learning behaviors (Merrill, 1983). A remember task entails only recall of facts, procedures, etc., with no application of the knowledge. At the use level, learners either use (apply) knowledge (at various levels) with or without an aid. As a design tool, the remember-use distinction is useful. Conceptually, however, the use level of learning tasks should be sub-divided, perhaps into convergent and divergent production and evaluation of responses (after Guilford, 1967) or some other more explicitly defined hierarchy of behaviors that reflects the ways in which learners process information. The point is, though, that learning *tasks* are separate from the *content being* processed.

I.2.2 Content Level

Content level refers to the type of information being processed, which in turn determines the type of knowledge constructions that are necessary for acquisition. Component display theory (Merrill, 1983) distinguishes between facts, concepts, procedures, rules, and principles as representing the overwhelming majority of content encountered in schools. Whether or not we use this taxonomy of content or other, more popular taxonomies (e.g., Gagné, 1977), we certainly need to define the type of information which is being processed in order to conduct a proper task analysis. So, the level of interaction between the learner and the learning system is largely a function of the task and content of the information being processed by the learner. Historically, too many interactive learning technologies (programmed instruction and its manifestations in CBI) require only recall of facts and concepts. In order for interactive learning systems to be useful, they need to replicate the task and content requirements of real-world tasks. Excellent examples of this are the aircraft pilot simulators and knowledge-based artificial intelligence programs that simulate decision-making processes, such as MYCIN. The latter system simulates diagnostic decisions for physicians. The entire instructional design process focuses on making instruction and evaluation procedures consistent with the nature of the task to be learned (*viz.*, medical diagnosis) rather than rote drill and practice.

I.3 Level of Processing

Closely associated with task analysis is the somewhat broader notion of levels of processing. The levels of processing hypothesis (Craik and Lockhart, 1972) provides an alternative framework for memory research. However, it has been generalized to areas of instructional design. Essentially, the hypothesis claims that as information is processed on subsequently deeper levels after perception and recognition, it involves a greater degree of semantic and cognitive analysis, that is, elaboration of the information based upon prior experience. Obviously, the deeper the level of processing, the more resistant will be the memory trace. Unlike hierarchical behavioral analysis, in which behaviors are classed as discrete types of behavior, the levels of processing hypothesis looks at processing as a continuum from shallow to deep processing. In designing interactive lessons, designers should be aware of the level of processing required by the learner responses. Generally, the deeper the level of processing represented by the responses, the more meaningful and resistant will be the learning. Hall (1983) related this principle to the design of questions embedded in computer-based courseware, although he employed Bloom's taxonomy rather than a strict levels of processing analysis. Jonassen (1984) described a number of learning strategy-oriented responses that could be embedded in courseware to raise (deepen) the level of processing of material.

The principle is simple: in designing interactive lessons, you want to encourage the learner to access prior learning and relate the new material more consistently to it. Learning is a constructive process. Interactive instructional designs should encourage the learner to construct more elaborate mental representations, rather than reflexively responding to information on a screen.

I.4 Type of Interactive Program

At a more general level, we can class the program as to its type: drill-and-practice, tutorial, problem-solving, simulation, or mixed-initiative, knowledge-based programs. While these classes are most frequently used to classify computer courseware, they can apply with equal accuracy to print-based or audio-visual materials. The type of program generally constrains the type of interactions involved. Drill-and-practice designs generally involve the most shallow processing of information. Tutorial designs frequently impose a set of constructs on the learner, requiring the learner to process the information at a deeper level. However, the opportunity to relate information to prior knowledge is not always evident, as it should be. Problem-solving designs require learners to apply rules and principles they have previously acquired to new instances, using the instructional technology as a tool for performing the operations. Simulations are distinguished from problem-solving in that the former require divergent responses as opposed to the usually convergent processing involved with the latter. The most meaningful interactive designs are known as intelligent systems—mixed-initiative, knowledge-based systems. The ability of these designs to conduct meaningful dialogue are far greater than that of any of the other designs, because the superior parsing ability of such programs enables natural language processing—the easiest for the learner to relate to prior knowledge. More importantly, these systems are capable of *learning from the learner* as well as guiding and directing the interaction based upon a pre-determined algorithm. While such systems are most frequently computer-based, it is possible to retro-fit some of these designs to the human-to-human interactions which they are trying to simulate.

I.5 Level of Intelligence of Design

Related to the previous discussion, the most abstract way of analyzing the type of technology-learner interaction is based upon the level of intelligence of the system. Traditional interactive designs are frame-oriented. The program presents a finite amount of information to which the learner responds and then presents feedback (on various levels) the learner. The learner then proceeds to the next frame or may be branched ahead or back to another frame in the program. The point is that the knowledge base is broken down into discrete units or chunks with which the learners interact one at a time. The shallowness of the processing involved in these designs results in part from the preclusion of integrative processing. If each frame is discrete, then there is little opportunity to integrate the knowledge in the program.

The next highest level of intelligence represented in interactive designs occurs in systems that include management components. Management systems range from simple collection and printing out of student responses to collection, analysis, diagnosis, and prescription of learning sequences. While the sophistication of these systems can be great, the level of their intelligence is limited by the fact that all of the capabilities are based upon immutable algorithms that are part of the design. Node-link systems (e.g., Denenberg, 1980) look more intelligent because their organization is based on subject matter structure. They access information (nodes) through various propositional links, so they look like a human knowledge base. However, it is only with knowledge-based, artificial intelligence systems that you acquire the ability of the system to acquire information from the learner and alter its own database. Human tutors can obviously do this (they don't necessarily).

The level of intelligence represented in an interactive design may be the most abstract level of analysis, but it is increasingly becoming the most important. The use of sophisticated hard technologies to deliver frame-oriented instruction that requires only shallow processing of information is becoming a more obvious waste of technology. The soft design technologies are increasingly accommodating cognitive principles of instruction, so the design of hard technology-based systems should increasingly reflect this orientation.

Adaptive Lesson Designs

Adaptive capabilities of instructional designs are those which enable the delivery of alternative instructional sequences based upon a variety of learner, content, or situational characteristics. Adaptive designs can be implemented in two ways. First, during delivery of instruction, the instructional program or technology may assess some characteristic(s) of the learner's performance or ability to perform, his/her preferences for mode of learning, or some curricular or personal need, and then adapt the sequence, strategy, or type of instruction presented to the learner based upon algorithms related to the characteristics assessed. Second, adaptive instructional programs can also adapt internally, based upon the nature of the task or information which is being presented to the learner at any time. So the adaptation of information may be based upon external information collected by the technology, or it may be programmed into the instructional presentation prior to delivery. Both approaches involve adaptation of the lesson strategy or content. In the first case, it represents adaptation to the learner; in the second, it represents adaptation to the information.

Such adaptations are usually implied as part of interactive instructional designs, because even the lowest levels of interaction, such as that in linear programmed instruction, minimally adapt to the learner, most often in terms of the pacing of instruction. It is possible, I suppose, to design programs which require learner involvement that would not adapt to the learner in any way. For instance, linear sequenced, externally-paced presentations, such as programmed audiovisual presentations that present information, insert a question, and then provide knowledge of results, don't adapt the sequence or style of presentation. Systems theory might argue that the feedback itself is implicitly adaptive, but in these presentations, there is no alteration of the presentation as such. So, the nature of adaptations in an instructional program may be controlled by external contingencies or by internal characteristics of the content. I shall summarize briefly different examples of each class of adaptation.

II.1 External Adaptation

II.1.1 Branching on Performance

The lowest level of adaptation occurs in instructional programs which branch on learner performance, such as in the various forms of branching programmed instruction. The simplest type of adaptation branches the learner to a remedial frame as a result of an error, and then sends him or her back to the original frame to answer the question. This form of branching is simplistic, because often the original question is a two-option branch, one of which leads to the remedial frame and the other which sends the learner on through the program. Even the least capable of learners can deduce that if one option in a two-option question is incorrect, the other *must* be correct, without processing the information presented in the frame.

Some allegedly more adaptive programs will merely branch learners to an earlier point in the program to repeat the sequence they did not comprehend when first processing it. In branching programmed instruction, the learner in principle must comprehend the information before answering a question, which is a confirmation of that comprehension, rather than the response portion of an S-R connection. Simple branching sequences, such as these, generally produce processing just as shallow as linear programming, that is, processing that does not require comprehension by the learner. More elaborate branching sequences can properly adapt instruction to individual needs.

Remedial loops based upon multiple branches, each of which anticipates the nature of the processing errors committed by the learner, are more truly adaptive. These and other forms of branching may diagnose learner weaknesses or strengths and branch them accordingly to appropriate remediation or to a point further on in the program which will adequately challenge them.

Without redundantly reviewing the history of programmed instruction, I simply want to point out that materials based on branching designs (a popular approach still) represent a form of adaptation designed to accommodate the program to learners' abilities to comprehend the material presented in it, that is, branching on performance. Most branching programs, however, do not truly adapt to learners, because programming truly adaptive sequences is more difficult, requiring a more elaborate information processing analysis of the task. Even the best branching designs are severely limited by the inherent restriction of frame-based instruction. The dialogue and the nature of the learner-technology interaction are constrained by having to prescribe the exact content of each and every possible interaction, and to program it into the material. So, adapting instruction based upon learner performance (as reflecting comprehension) is severely constrained by the use of a traditional frame-based form of branching instruction.

II.1.2 Diagnosing Prior Learning

Some instructional designs call for early assessment of the learner's background knowledge or ability level. The designs will analyze the results of a diagnostic test and then use a predefined algorithm to place the learner at a level in the instructional sequence in which he or she will be able to achieve mastery of the material. The idea of such placement is to challenge the learner but not frustrate him/her. Many individualized learning systems that were popular in the schools during the seventies were based upon this model. Usually, they were arranged in a sequence of objectives supported by a variety of instructional materials. The learners would be diagnosed and placed at a point in the system at which they could achieve at a prescribed level of performance (e.g., 70 percent). Such diagnosis can also be a part of self-contained materials. The design process consists generally of identifying a sequence of cumulative learning skills, breaking them down into an objectives-based curriculum, and then designing a diagnostic instrument for placing the learner in that sequence. Many of the microcomputer skills series so popular today are based on this adaptive model. It is practical and therefore popular in agencies, such as schools, which value practicality.

II.1.3 Accommodating to Learner Characteristics

Based upon a prodigious body of research, we can conclude that learners' abilities to process and acquire different types of information, that is, to think in different ways, is mediated by a diverse variety of individual differences. Based upon a slightly less prodigious body of research, on aptitude-treatment interactions (Cronbach and Snow, 1979), we can also conclude that an individual's ability to perform certain skills or acquire certain information may be mediated by the nature of the instructional intervention or treatment. The logical implication is that we can accommodate individual differences in processing ability by altering the nature of the learner interaction in order to enable individuals with different processing capabilities to learn. The most obvious and productive learner characteristic on which to adapt our instruction is nature of the instructional treatment. For instance, we might reorganize, resequence, or redesign the way information is presented.

Intellectual ability usually accounts for the greatest amount of variance in learning. Allen (1975) provides numerous recommendations as to how to structure and design materials for low, middle, and high ability learners.

Related to intelligence or ability is the developmental variable of age. Numerous stage theories of development (e.g., Piaget, Bruner, Kohlberg) assert that intellectual processing and learning skills are age-specific. That is, as learners grow and develop, they acquire more sophisticated information processing abilities. The obvious, adaptive implication of these theories is that learning tasks should be geared to the developmental capabilities of the learners. Once we know the level of processing the learner is capable of (e.g., concrete operational), we can prescribe instructional tasks that challenge but don't frustrate the learner (e.g., tasks requiring concrete, not abstract, logic).

We may also provide instructional programs that adapt the treatment on the basis of different cognitive styles of learners. For instance, the cognitive style of leveling/sharpening describes the tendencies of learners to integrate specific attributes of memory into a construct (levellers) or to recall information as discrete memories (sharpeners). If our task required listing in sequence, we might embed mnemonic strategies in the treatment we give to sharpeners. If the task required comprehension, we might use an organizer or an imagery strategy in the material. Numerous studies have examined alternative treatments for such cognitive style variables as field dependence-independence, visual-haptic style, cognitive tempo (impulsivity-reflectivity), breadth of scanning, conceptual style, and many others. For some of the implications of cognitive styles for instructional design, see Ausburn and Ausburn (1978).

There are numerous other individual differences to which instructional treatments may be adapted. For instance, motivation plays a major role in learning. For learners who are not intrinsically motivated, behavior modification techniques or other reward contingencies may be programmed into materials. Some instructional microcomputer materials reward effort with the opportunity to play an arcade style game after so many correct responses. We certainly could adapt instructional materials to the social context of the learner. Whenever cultural bias might be a problem, materials with culturally relevant instantiation could be included for each different group who may use the material. The instructional algorithms would be the same; only the examples would change. These are but a few of the individual differences to which it is possible to adapt instruction.

II.1.4 Learner Control of Instruction

The simplest type of adaptive design to implement organizes the presentation, identifies a variety of options, and then allows the learner the opportunity to select the scope, sequence, method, style, or amount of instruction. Rather than having the program control the nature of instruction, Merrill (1975) recommended that designers go beyond aptitude-treatment interactions by allowing learners to control their own instruction. At the very least, it is important to the design of interactive programs that learners be able to choose at will to exit the program, review sections, get help, or select sequencing or testing options. Learner controlled instruction may allow learners also to choose what information is to be displayed, the sequence in which it is to be displayed, the number of examples of concepts presented, the number of practice questions they want to complete, or the modality (sound, graphics, etc.) of the presentation.

Most criteria for evaluating microcomputer courseware now recommend that learners be given maximum control of instruction, not only in terms of the pacing of the presentation but also what and how it is being displayed. Putting learners in control of instruction seems a fair and mature approach to learning. However, some caution needs to be exercised. Clark (1982), for instance, concluded after reviewing a lot of research, that learners frequently prefer and therefore choose the method of instruction from which they learn the *least*. This is especially true as the task becomes more difficult and the content structure becomes more complex (Tennyson, 1980). Certainly, some control of instruction by the learner should be encouraged by instructional systems; however, that control will succeed only if learners are capable and willing to monitor their own comprehension. In order to do that, the instructional program needs to include instructional interactions which use meta-cognitive strategies. Learners should be encouraged to approach learning maturely. To do so, they need the meta-cognitive skills necessary for effectively controlling their instruction. These may be practiced, and to a limited extent, trained by the instructional materials (Jonassen, 1985; Pace, 1985).

II.1.5 Knowledge-Based Systems

The most meaningful form of instructional adaptations occur in a dialogue in which the teacher adapts the logic, sequence, and style of instruction based upon an understanding of the structure of the knowledge to be acquired and the ability to perceive how the learner is trying to acquire the information—a true Socratic dialogue.

The only instructional technologies which approach these capabilities are referred to as "intelligent tutor systems." Unlike frame-based systems, which prescribe the user-system dialogue and the logic which controls it, intelligent systems, also referred to as information-structure-oriented systems, use a network of information to generate text or questions and respond to student queries (Carbonell, 1970). So, another distinction between frame-based and information-structured instructional systems is that the latter are mixed-initiative. That is, the learner may initiate the dialogue by asking a question, rather than merely being able to answer questions prescribed by the program.

Intelligent systems must contain representations of knowledge (semantic content as well as procedural skills), a formal grammar for conducting the dialogue, and a production system, which consists of the rules which drive the system (Bregar and Farley, 1980).

A detailed discussion of intelligent systems is beyond the scope of this article, but it should be noted that the most meaningful adaptations are those that are not based on a prescribed sequence of interactions. Adaptations should rather be based on an understanding of what the learner is thinking. In order to do this, instructional technologies need to be oriented by a knowledge base capable of representing what the learner is in fact thinking during the process of instruction.

II.1.6 Curricular Adaptations

At the broadest level of external adaptation, instructional systems may track learners according to curricular prescriptions. Such adaptations have been referred to as utilization-related (those that focus on social, personal, or career needs), often those related to certification needs. At this level, such systems may contain elements of all of the other types of adaptation which are learning-related.

From a learning perspective, they may appear only as the arbitrary assignment of materials based upon a prescribed curriculum. Curricula are usually arbitrary, based on the arbitrary assignment of content, rather than any assessment of the information needs of the learners, not to mention the assessment of the ability of the learners to comprehend the content.

Learning-related systems would be more responsive, however, to those aspects. Utilization-related systems frequently track learners into grade level or subject matter slots irrespective of the learning needs or abilities of the learners, the nature of the tasks involved, or the relationship between them. Curricular approaches often represent a very shallow form of adaptation that is based on arbitrary exit criteria.

II.2 Internal Adaptations

Just as aptitude-treatment interaction research assumes that no treatment is appropriate for all learners, I contend that no treatment is necessarily appropriate for all *content* (Jonassen, 1982). There are different types of content (facts, concepts, principles, procedures, and problems). Different levels of content require different information processing operations in order to remember, comprehend, or use them.

The way that information is structured and coded will surely affect the way it is processed. Therefore, different content should be treated differently. That is, different content requires different types of information processing tasks. Since instructional treatments that model the processing requirements of the task result in greater learning, instructional programs should adapt to conditions internal to the program, i.e., the nature of the content presented and the tasks required. I shall review briefly some of the types of internal, content-, and task-oriented adaptations.

II.2.1 Task Requirements

Task requirements may vary as a function of the content, as stated above, or as a function of the stated objectives for the program. Learners should receive the treatment that best simulates the processing requirements which result in the most efficient learning. For instance, if our objective is for the learner to apply a cause-effect principle, it should be presented in a way that models the relationship between the cause and effect. One of the most important roles of the designer then becomes information processing task analysis (Merrill, 1978). The designs that result from the analysis should seek to model the processes identified by the analysis as required by the objective.

II.2.2 Content Sequencing

The sequence of the presentation may be altered to accommodate differences in task or content requirements. An obvious example would be alternatively sequencing information inductively or deductively. The inductive or discovery approach employs an EGRULE sequence, where examples are presented and learners induce the rule. In the deductive, expository, or RULEG sequence, the generality is followed by examples. If the goal of instruction is classification, the latter should be used. When you want learners to transfer rules to novel situations, it is generally recommended that the former sequence be used. Similar differences would result in teaching relational versus analytical

concepts. Content varies; so should sequences for presenting it.

II.2.3 Content Structure

The nature of an instructional program may vary based upon the structure or organization of the content you are presenting. That organization can be signaled explicitly to make it easier to comprehend. The structure of the content may also be reflected in the structure and sequence of the presentation. For instance, if the elaboration theory of instruction were employed, the learner would first be presented with an overview, which would then be elaborated sequentially and iteratively until a sufficient level of detail were achieved. This process of synthesis-analysis-summary is designed to make explicit the organization of subject matter. Different organizations or contexts may be activated during instruction by using a variety of advance organizers. The point of each of these sequences is that comprehension of content improves if learners are more aware of the structure of the knowledge being studied.

II.2.4 Curriculum Development

Curricular adaptations may also be internal to the content, reflecting primarily the structure of the discipline rather than the nature of the learner's performance. The attempt in structuring content at this level is to represent things the way they are. The different classes of content structure have been conceptualized to include world-related (real-world perspective), concept-related (logical structure of the conceptual world), and inquiry-related (discovery process). They represent the broadest perspective from which we can view the adaptation process. However, such a view has produced some very excellent instructional systems, such as BSCS biology and PSSC physics. Disciplines such as biology and physics may be structured in such a way (structured inquiry, in these cases) as to render the instruction more effective.

Summary

Interactive technologies possess two dimensions, the interactive dimension and the adaptive dimension. Often, these two are not distinguished. Yet they each add a different set of capabilities to interactive programs.

The interactive dimension describes the way in which learners interact with an instructional program. The nature of that interaction was described in terms of the task, the level of processing required to complete that task, and the context in which the instructional program is used.

The adaptive dimension describes the way in which instructional programs may adapt to either the needs of the learner or to the content it is presenting. An instructional technology may be designed to adapt to the information needs of the learner, his/her preferred style or mode of learning or ability to learn, or to the instructional context in which the learner is operating. Additionally, instructional technologies may be designed to adapt internally to nature of the task being learned, the structure of the content, or the curricular context in which the material is being acquired.

When designing interactive, adaptive learning systems, the designer has numerous design options available. According to the design selection model shown in Figure 1, the designer has at least 144 different ways of varying the presentation! Some of the most difficult design decisions will concern the matching of types of interactivity with adaptation, both internal and external. For the specific learners and the nature of the task being learned, do you match instructional characteristics to capitalize on learner strengths, to remediate or supplant learner deficiencies? Or do you intentionally challenge the learners to operate in levels of interaction and adaptation in which they are deficient so that they will acquire the skill? These are issues beyond the scope of this article but important to the ideas presented in it.

The purpose of this article is simply to classify the types of interaction and adaptation available to designers. This is important because the newer, computer-based technologies are so much more capable of accommodating elaborate interactive, adaptive designs.

Too much of the work in these technologies, such as interactive video, start with an analysis of the capabilities of the technology. Despite the awesome potential of these technologies, we must start with an analysis of the problem and develop solid instructional designs before considering the medium. The primary application of the taxonomy of interactive, adaptive lesson design described in this article is as a selection guide for choosing the appropriate technology. Once the type of interaction and adaptation most appropriate for an instructional problem are identified, the designer needs simply to use a checklist of the capabilities of various interactive technologies to assess the ones capable of delivering that type of design. Then other comparisons, such as cost-effectiveness, may be used to choose between alternative delivery systems. But the design process requires at least knowing the design possibilities available. ☐

References and Suggested Readings

Allen, W.H. Intellectual Abilities and Instructional Media Design. *A V Communication Review*, 1975, *23*, 139-170.

Ausburn, L.J., and Ausburn, F.B. Cognitive Styles: Some Information and Implications for Instructional Design. *Educational Communications and Technology Journal*, 1978, *26*, 337-354.

Barr, A., and Feigenbaum, E.A. *The Handbook of Artificial Intelligence*, Vol. 2. Stanford: HeurisTech Press, 1982.

Bregar, W.S., and Farley, A.M. Artificial Intelligence Approaches to Computer-Based Instruction. *Journal of Computer-Based Instruction*, 1980, 6(4), 106-114.

Carbonell, J.R. AI in CAI: An Artificial Intelligence Approach to Computer-Assisted Instruction. *IEEE Transactions on Man-Machine Systems*, 1970, MMS-11 (4), 190-202.

Cronbach, L.J., and Snow, R.E. *Aptitudes and Instructional Methods.* New York: Irvington, 1979.

Clark, R.M. Antagonism Between Achievement and Enjoyment in ATI Studies. *Educational Psychologist*, 1982, *17*, 92-101.

Cohen, V.B. Interactive Features in the Design of Videodisc Materials. *Educational Technology*, 1984, *24*(1), 16-20.

Craik, F., and Lockhart, R. Levels of Processing: A Framework for Memory Research. *Journal of Verbal Learning and Verbal Behavior*, 1972, *11*, 671-684.

Dayton, D. Inserted Post Questions and Learning from Slide Tape Presentations: Implications of the Mathemagenic Hypothesis. *A V Communication Review*, 1977, *25*, 125-146.

Denenberg, S.A. Using a Semantic Information Network to Develop Computer Literacy. *Journal of Computer-Based Instruction*, 1980, 7(2), 33-40.

Gagné, R. *The Conditions of Learning*, 3rd Ed. New York: Holt, Rinehart, and Winston, 1977.

Guilford, J.P. *The Nature of Human Intelligence.* New York: McGraw-Hill, 1967.

Hall, K.A. Content Structuring and Question Asking for Computer-Based Instruction. *Journal of Computer-Based Instruction*, 1983, *10*(1&2), 1-7.

Jonassen, D.H. Aptitude- vs. Content-Treatment Interactions: Implications for Instructional Design. *Journal of Instructional Development*, 1982, *5*(4), 15-27.

Jonassen, D.H. The Electronic Notebook: Embedding Learning Strategies in Courseware to Raise the Level of Processing. In B. Alloway and G. Mills (Eds.), *Aspects of Educational Technology*, Vol. 16. London: Kogan Page, 1984.

Jonassen, D.H. Generative Learning vs. Mathemagenic Control of Text Processing. In D.H. Jonassen (Ed.), *The Technology of Text: Principles for Structuring, Designing, and Displaying Text, Vol. 2.* Englewood Cliffs: Educational Technology Publications, 1985.

Kearsley, G. Instructional Design Considerations for CAI for the Deaf. Paper presented at the annual meeting of the Association for the Development of Computer-based Instructional System, Dallas, Texas, March 1-4, 1978.

Langdon, D. *Interactive Instructional Designs for Individualized Learning.* Englewood Cliffs: Educational Technology Publications, 1973.

Merrill, M.D. Learner Control: Beyond Aptitude/Treatment Interaction. *AV Communication Review*, 1975, *23*, 217-226.

Merrill, M.D., Component Display theory. In C.M. Reigeluth (Ed.), *Instructional Design Theories and Models: An Overview of Their Current Status*. Hillsdale, NJ: Lawrence Erlbaum Associates, 1983.

Merrill, P. Hierarchical and Information Processing Task Analysis: Comparison. *Journal of Instructional Development*, 1978, *1*(2), 25-40.

Meyer, B.J.F. Signaling the Structure of Text. In D.H. Jonassen (Ed.), *The Technology of Text: Principles for Structuring, Designing, and Displaying Text, Vol. 2*. Englewood Cliffs: Educational Technology Publications, 1985.

Pace, A.J. Learning to Learn Through Text Design: Can It Be Done? In D.H. Jonassen (Ed.), *The Technology of Text: Principles for Structuring, Designing, and Displaying Text, Vol. 2*. Englewood Cliffs: Educational Technology Publications, 1985.

Rothkopf, E.Z. The Concept of Mathemagenic Activities. *Review of Educational Research*, 1970, *40*, 325-336.

Tennyson, R.D. Instructional Control Strategies and Content Structure as Design Variables in Concept Acquisition Using Computer-Based Instruction. *Journal of Educational Psychology*, 1980.

Learning the ROPES of Instructional Design: Guidelines for Emerging Interactive Technologies

---

Simon Hooper and Michael J. Hannafin

The last decade has been marked by an exponential increase in the development of educational technology. From computer-based instruction (CBI), to computer-assisted interactive video (CAIV), to compact-disc read only memory (CD-ROM), to compact-disc interactive (CD-I), and now digital video interactive (DV-I), educators have available the most advanced toolbox ever created with which to design instruction.

Some perceive the growth of emerging instructional technologies as a challenge to create more effective software. Often this challenge is based on the flawed assumption that *more complex technology* is analogous with *better instruction*. A symptom of this assumption is a tendency to focus design on the technical attributes of new media. This technocentric perspective views *technology* as central to the learning process instead of the student.

Recently, Hannafin and Phillips (1987) developed a meta-model to classify the myriad of research and theory in learning and instruction according to instructional phases. We refer to this model as the ROPES of instruction; Retrieval, Orientation, Presentation, Encoding, and Sequence. Although designed specifically for the development of CAIV, the model is generic since it is based on psychological rather than technological theory and research. Hooper, Hannafin, and Phillips (1987) proposed a number of guidelines, based on the model, aimed specifically at the design of CAIV. The objective of this article is to extend the range of the model by presenting a number of empirically derived design guidelines for emerging interactive technologies. Each guideline is based upon research and theory in learning, instruction, and/or media development.

Simon Hooper is with the Instruction Support Center, and Michael J. Hannafin is with the Center for Research and Development in Education Computing, The Pennsylvania State University, University Park, Pennsylvania.

Retrieval

The goal of information retrieval permeates all phases of instruction. Design guidelines should, therefore, be considered within the context of how they relate to retrieval. The various phases of ROPES are designed to improve the retrievability of instructional content.

To retrieve information from long term memory (LTM) some cueing mechanism is required. The retrieval cue is generally activated in one of two ways: The cue may be produced automatically in response to a question, or a student may generate a cue either by searching existing cognitive structures for relevant information or by applying a strategy to generate cues. Guidelines in this section are designed to facilitate the generation of cues.

Embed strategies that facilitate meaningful learning. Since encoding and retrieval are "inextricably tied" (Klatzkey, 1980, p. 236) we must examine conditions of encoding likely to result in effective recall. In particular, since retrieval is influenced by how information is processed initially, methods that result in deep processing should be investigated.

Mayer (1984) described three levels at which lesson information may be learned. First, information is perceived. Perception may be facilitated by placing cues close to important text. For example, underlining or otherwise highlighting text is likely to improve perception. Second, information is organized. Organization can be enhanced by any of a number of techniques that highlight the structure of information. For example, headings may help students to link important ideas by forming internal connections. Organization within short-term memory facilitates retrieval by improving the availability of cues in LTM. Third, information is integrated resulting in meaningful encoding. Integration may be facilitated through the use of strategies that build external connections between organized information and existing knowledge structures in LTM. For example, advance organizers and chapter summaries help students to link information within existing cognitive structures (DiVesta and Rieber, in press).

Relate instructional content to students' prior experience. Information that cannot be immediately recalled may be constructed by examining the contents of related schemata. Schemata are constructed, refined, and otherwise modified by combining new with existing information during initial learning. Generally, the closer the link between new and existing information the better instantiated the schema and the more effective the retrieval process. Conversely, trying to teach meaningless information is likely to prove fruitless since the learner has no schema within which it can be subsumed.

One way to encourage meaningful schema development is by personalizing information based on a pool of individually relevant data. Anand and Ross (1987) found that when personal information was embedded in math problems, students performed well on tests of retention, problem solving, and near transfer. Presumably, they found it easier to integrate the new information within existing schemata.

Orientation

Orienting activities help to prepare learners for instruction by retrieving relevant information from LTM to be encoded with new information. In general, these devices may be either explicit, such as behavioral objectives that help to acquaint the learner with a specific task, or implicit such as advance organizers, that help the learner to build an internal framework for the learning task. Orientation design decisions must be based upon intended cognitive outcomes and motivational issues.

Orienting activities must be aligned thematically with the intended learning task. With the rapid advancements in emerging instructional technologies, there is a temptation to emphasize the capabilities of the medium. Lessons are often dramatic in their look and sound. However, these capabilities will often impede rather than enhance learning if designed carelessly. Congruence between orienting activities and the learning task is essential in order to direct attention to the relevant features of a lesson.

Explicit orienting activities should enhance the learning of specific information, while implicit activities should support higher level learning tasks. Although specific behavioral objectives facilitate the learning of information directly related to objectives, they also tend to restrict the learning of incidental information (Anderson, 1970). On the other hand, implicit orienting activities, that focus on how information is learned, facilitate the building of frameworks that help students to integrate information (Ausubel, 1968). For example, in certain cases factual information may be better learned through the use of specific behavioral objectives. However, higher level learning requires meaningfulness and understanding, and may be better mediated by the use of implicit orienting activities such as advance organizers (Hannafin and Hughes, 1986).

Design orienting activities that enhance and/or manipulate motivation. The affective orienting activity is designed to heighten arousal and thus increase motivation. While increasing arousal alone is unlikely to improve learning, it is likely to make the learner more responsive.

Educators should consider the potential of each of several motivational orienting activities when designing instruction: Incentives motivate by offering intrinsic or extrinsic rewards. "Grabbers" have the effect of both gaining attention and orienting the learner to the impending instruction; impact scenarios relay lesson objectives through the use of vivid descriptors, such as the danger of crossing the road between parked cars; finally, group contingency rewards can motivate learners by requiring team members to be responsible for each others' progress (Hannafin, 1987).

Presentation

Recent innovations offer the instructional designer the potential to employ a wealth of advanced technology to facilitate instruction. However, when perceived as "easy," students may not process information very successfully. Designers must take great care when using technology to ensure that learners process information adequately.

Vary video and non-video techniques to distribute emphasis purposefully and systematically. Research suggests that, when stories are read aloud, students tend to process information to greater depths than students who receive the same stories via TV (Meringoff, 1980). Similarly, Salomon (1984) found that children perceived TV as easier than print, but that the increased mental effort required from print resulted in improving learning. Consequently, one potential pitfall of TV and video is that students will fail to process adequately information. When video is employed, designers should take steps to highlight important information through the varied capabilities of technology, causing students to process those aspects of the lesson more deeply and completely. For example, windowing techniques, that involve masking non-essential information, encourage the student to focus on important lesson content.

Color has the potential to attract to as well as detract from instructional messages, and must be employed judiciously. Clearly, color enhances the potential to highlight information and make visual presentations motivational. However, since color may increase the amount of information to be encoded, conscious and cautious decisions should be made concerning its use. In general, designers should recognize that the effectiveness of color to influence selective perception is likely to diminish with increased use (Kanner, 1968). Color may be most effective when used to exagerate differences and attract attention to key lesson features.

Image quality, realism, and detail are only essential when the learning task calls for those particular attributes, but may support affective

consequences of the lesson. New instructional technologies offer the potential for the high-fidelity natural presentation of visual and aural images. However, attributes such as quality, realism, and detail often increase the available stimuli, and introduce extraneous and irrelevant information to the task (Dwyer, 1978). For example, a simple line drawing representing the heart may be more effective for illustrating blood flow than an intricate, high quality, and highly realistic heart model.

Combine modes of presentation to enhance depth of processing. One theory suggests that the combined presentation of different forms of media is likely to have an additive effect on learning if the different media are complementary (Paivio, 1979). Thus, for example, media that employ both print and video are likely to result in deeper processing than a medium that employs just print—assuming, of course, that the content of both media is closely related, and that the presentation stimuli do not compete for learner attention and effort.

Encoding

Encoding requires new information to be organized within an existing cognitive structure. Encoding guidelines are designed to facilitate transfer into LTM based upon factors that affect the integration of new with existing information.

Incorporate prompts that facilitate comprehension monitoring. Students of almost any age can be taught learning tactics or strategies (Derry and Murphy, 1986) such as note-taking and reading techniques. However, tactical knowledge alone is unlikely to improve learning, without an executive control mechanism to regulate application.

Metacognition involves the recognition of learning difficulties and regulation of learning strategies. Metacognitive behavior may be fostered in instruction through the insertion of prompts that remind the learner to monitor comprehension and suggest learning strategies that may facilitate learning. For example, students may be advised to employ mnemonic aids when learning lists of factual information.

Utilize strategies to facilitate processing: Wait time enhances higher cognitive thought; distributed practice and other forms of guidance reduce cognitive load. Designers must remain cognizant of the limited capacity of short term memory, and the corresponding implications for instruction: Instruction that requires too much information to be processed in too little time is likely to result in an inefficient or even dysfunctional learner (Travers, 1982). Consequently, information must be presented at a rate that matches the processing capabilities of the learner.

Designers of interactive technologies may influence processing by manipulating wait time. Wait time is the amount of time allowed for a response. While short wait time may be an advantage for learning factual information, learning higher cognitive information requires greater time to allow adequate processing (Tobin, 1987).

Practice in the form of various types of questions has long been valued as an instructional strategy. The distribution of practice activities not only varies lesson activities and permits assessment of learning, but serves a number of psychological functions as well, by reducing demand on short term memory and facilitating recall (Salisbury, Richards, and Klein, 1985).

Feedback should identify the steps involved in the correct solution. Webb (1983) reported widely different reactions to different types of feedback. When examined as a whole, feedback had little influence on learning. However, when differentiating between types of feedback, Webb found that receiving explanations of the solution process was correlated positively with achievement. Further, receiving terminal responses (responses without explanations) or ignoring requests for help, was detrimental to success. Although Webb's work related specifically to cooperative learning, the findings seem likely to transfer to individual instruction as well.

Sequence

The issue of lesson sequence is often viewed simplistically as a question of how much autonomy should be given to the learner. However, the primary issue concerning lesson sequencing should not concern whether students are given lesson control, but rather what *type* of control students are given. Clearly, control decisions concerning whether a student may skip sections of a lesson or terminate instruction are of far greater consequence than support decisions concerning whether a student may review instruction or receive additional help. Correspondingly, designers should provide learner controlled on-line support as a standard feature.

Allow the learner to determine lesson sequence when content is familiar or poses little cognitive difficulty; provide guidance when learner control is selected. One of the goals of education is to produce independent learners capable of managing their own learning environments; learner productivity improves with independence (Reigeluth and Stein, 1983). Computer-based education appears to promote the goal of the independent learner by allowing learners to determine their own sequence through a lesson. However, research suggests many students are incapable of reaping the benefits of certain types of lesson control. Many are unsuccessful when given the opportunity to control the sequence of instruction, the number of questions to answer, and the length of presentation (Tennyson, Christenson, and Park, 1984).

Learner control of such features is often ineffective because many learners are unable to determine how much instruction is adequate. However, when given notice of progress toward an objective and advice on how to proceed, students perform comparably under both guided and computer control (Tennyson and Buttrey, 1980). Thus, designers should be sensitive to the need for "expert" advice during learner control of lessons.

As an alternative to linear designs, use adaptive designs to match instruction to individual needs. Linear lessons, that guide all learners through identical instruction regardless of ability or prior knowledge, ignore the potential of the computer to individualize instruction. Further, linear designs may be counterproductive: More experienced learners may have learning styles that conflict with imposed learning strategies. Nevertheless, under some circumstances, lessons may require all students to demonstrate some pre-defined level of performance. Adaptive designs can guarantee performance, but still remain sensitive to individual learning needs. Thus students who demonstrate a high level of proficiency may test-out of unnecessary instruction.

Branching should be sensitive to both macro and micro performance indicators. Adaptive instruction is often based upon macro performance indicators such as individual performance on questions embedded in text. Performance on these questions influences the number of questions and examples presented to each student. The less obvious micro indicators, such as adapting display time to individual needs, are often ignored. For example, a student who makes a disproportionate number of errors in a lesson, coupled with a decrease in display time, may indicate need for increased motivation rather than a need for more questions or examples.

Summary

This article is not intended as a definitive list of design guidelines. Rather, its function is to illustrate how existing theory and research, in learning and instruction, may serve as a basis for design guidelines. At present, guidelines often focus on technological issues rather than paying heed to issues of pedagogy: Instead of "reinventing the wheel" we have attempted to extrapolate well defined principles of cognition and apply them to the design of lessons featuring emerging instructional technologies. □

References

Anand, P.G., and Ross, S.M. Using Computer-Assisted Instruction to Personalize Arithmetic Materials for Elementary School Children. *Journal of Educational Psychology*, 1987, *79*, 72-78.

Anderson, R.C. Control of Student Mediating Responses During Verbal Learning and Instruction. *Review of Educational Research*, 1979, *40*, 349-369.

Ausubel, D.P. *Educational Psychology: A Cognitive View.* New York: Holt, Rinehart, and Winston, 1968.

Derry, S.J., and Murphy, D.A. Designing Systems that Train Learning Ability: From Theory to Practice. *Review of Educational Research*, 1986, *56*, 1-39.

DiVesta, F.J., and Rieber, L.P. Characteristics of Cognitive Instructional Design. *Educational Communication and Technology Journal*, in press.

Dwyer, F.M. *Strategies for Improving Visual Learning.* State College, PA: Learning Services, 1978.

Hannafin, M.J. *Motivational Aspects of Lesson Orientation during CBI.* Presented at the annual Meeting of the American Educational Research Association, Washington, D.C., 1987.

Hannafin, M.J., and Hughes, C.W. A Framework for Incorporating Orienting Activities in Computer-Based Interactive Video. *Instructional Science*, 1986, *15*, 239-255.

Hannafin, M.J., and Phillips, T.L. Perspectives in the Design of Interactive Video: Beyond Tape Versus Disc. *Journal of Research and Development in Education*, 1987, *21*, 44-60.

Hooper, S., Hannafin, M.J., and Phillips, T.L. *Psychologically-Based Guidelines for the Design of Instructional Interactive Video.* Submitted for publication, 1987.

Kanner, J.J. *The Instructional Effectiveness of Color in Television: A Review of the Evidence.* Stanford: Stanford University (ERIC; ED 015-675), 1968.

Klatzkey, R.L. *Human Memory* (2nd ed.). San Francisco: W.H. Freeman & Co., 1980.

Mayer, R.E. Aids to Text Comprehension. *Educational Psychologist*, 1984, *19*, 30-42.

Meringoff, L.K. Influence of the Medium on Children's Story Apprehension. *Journal of Educational Psychology*, 1980, *72*, 240-249.

Paivio, A. *Imagery and Verbal Processes.* Hillsdale, NJ: Lawrence Erlbaum Associates, 1979.

Reigeluth, C.M., and Stein, F.S. The Elaboration Theory of Instruction. In C.M. Reigeluth (Ed.) *Instructional Design Theories and Models: An Overview of Their Current Status.* Hillsdale, NJ: Lawrence Erlbaum Associates, 1983.

Salisbury, D.F., Richards, B.F., and Klein, J.D. Designing Practice: A Review of Prescriptions and Recommendations from Instructional Design Theories. *Journal of Instructional Development*, 1985, *8*(4), 9-19.

Salomon, G. Television Is "Easy" and Print Is "Tough": The Differential Investment of Mental Effort in Learning as a Function of Perceptions and Attributions. *Journal of Educational Psychology*, 1984, *76*, 647-658.

Tennyson, R.D., and Buttrey, T. Advisement and Management Strategies as Design Variables in Computer-Assisted Instruction. *Educational Communications and Technology Journal*, 1980, *28*, 169-176.

Tennyson, R.D., Christensen, D.L., and Park, S.I. The Minnesota Adaptive Instructional System: An Intelligent CBI System. *Journal of Computer-Based Instruction*, 1984, *11*, 2-13.

Tobin, K. The Role of Wait Time in Higher Cognitive Level Learning. *Review of Educational Research*, 1987, *57*, 69-95.

Webb, N.M. Predicting Learning from Student Interaction: Defining the Interaction Variables. *Educational Psychologist*, 1983, *18*, 33-41.

Improving the Meaningfulness of Interactive Dialogue in Computer Courseware

George R. McMeen and Shane Templeton

"Language, for all its kingly role, is in some sense a superficial embroidery upon deeper processes of consciousness. . . ." (Whorf, 1956, p. 239).

Introduction

The presentation of ideas and the learning of concepts usually involves discourse, as in many interactive tutorial dialogues. Even rhetoric may be present in the display of information, as sentences are generated to involve the learner and introduce meaningful material to be learned. Computer dialogue establishes a series of acts or events in which the program's organization and structure reinforce the dialogue's persuasiveness and educative quality. Acts or events within this context have a syntactical and even rhetorical organization and structure that affect the program's persuasiveness and instructiveness as well as its ability to represent or reconstruct reality. Dialogue is frequently related to pictorial information, making use of the computer's capacity for graphic displays. Improvements in interactive dialogue need to be explored in terms of their relationship to pictorial meaning.

How can interactive dialogue in computer-assisted instructional programs be improved in a way that relates content to the deeper processes of consciousness? How can the instructional designer better structure interactional dialogue for the learner? Besides the general necessity of cleaning up what may be a verbal clutter of writing problems that computer programmers/authors may bring to the computer, the answer to these questions lies in considering how to make learning at the computer terminal more meaningful. It may also encompass the relationship of verbal and pictorial information. Context should be considered in the design and development of verbal information within any computer program, since it may be used to awaken cognitive schemata, or structures, in memory, where it plays an important role in the development of "skeletal" schemata and in the learning of meaningful verbal information. This is particularly important when new information is presented to help the learner assimilate new knowledge (Winn, 1981, p. 8). Moreover, the design of interactive dialogue is important if concepts are to take their hierarchical order in memory.

Research in both memory and the reading process has provided an idea of how better to structure the relationships within and between ideas or "propositions" in texts. This knowledge is being used in some computer programs to assist the writer (*e.g.*, MacDonald, Frase, Gingrich, and Keenan, 1982), but as yet it has not been applied on a broader scale to the preparation of instructional software. Coherent, engaging dialogue can no doubt be written more or less intuitively by some, but rather than trusting to the "we know what it is when we see it" approach, we should be guided instead by efforts to support readers' comprehension through effective text design. These efforts have been described by Glynn and Britton (1984). Our purpose here is to suggest the possible role of meaningful verbal and pictorial information in interactive dialogue at the computer terminal.

What Is Meaningful?

Different interpretations of what is meaningful in learning may be found in verbal and pictorial research. For instance, meaningful verbal learning has sometimes been associated with the subsumption or hierarchical arrangement of concepts in the learner's memory in relation to their level of inclusiveness. The view that more inclusive concepts subsume less inclusive ones may be traced to the theory of meaningful reception learning (Ausubel, 1968), which presupposes that inclusive concepts and their more general aspects are retained in cognitive structure, while specific details of less importance are forgotten.

In studies of pictorial learning, meaningfulness has been perceived as the association value regarding abstraction and realism in pictorial recognition (Koen, 1969). It has also been interpreted as familiarity found in recognition accuracy of levels of abstraction and realism in word-picture associations (Franzwa, 1973).

Investigations such as these have usually focused upon key elements in meaningfulness, but no single definition adequately describes the full range of

George R. McMeen is Director, INSTR Associates, Instructional Computing and Consulting, and Associate Professor, Department of Curriculum and Instruction, College of Education, University of Nevada-Reno, Reno, Nevada. **Shane Templeton** is Associate Professor in the same department.

possible definitions for meaningful verbal and pictorial learning and retention. Descriptions of pictorial meaningfulness are not easily developed, and this fact may account partly for the lack of a more complete definition. Another reason is that determining appropriate schemata in cognitive structure is difficult. Even schemata may change as they receive and assimilate new information by being modified in a reciprocal way (Anderson, 1977; Neisser, 1976; Piaget, 1967).

Theories of meaningful reception learning (Anderson, Spiro, and Montague, 1977; Ausubel, 1968; Lindsay and Norman, 1977; Piaget and Inhelder, 1969) postulate the existence of a cognitive structure that is hierarchically organized in terms of conceptual schemata which subsume less inclusive concepts. As a learner acquires new concepts from written verbal material, information is linked to existing schemata in cognitive structure. New concepts are assimilated through a process of subsumption, which involves the assimilation of new information in memory. Subsumption may be affected by the availability of specific cognitive elements or schemata for the anchorage of new concepts. The level of inclusiveness and generality of these concepts has an effect upon the manner in which they are subsumed or assimilated. For instance, more inclusive concepts are more general than less inclusive ones, and they subsume more specific concepts as material is acquired. Thus, one concept may retain another. This process is called progressive differentiation. However, the details associated with less inclusive concepts are discarded and forgotten in a process called obliteration. As the learner progressively differentiates new material within memory, this information is organized under a pre-existing ideational framework of previously learned concepts. In the case of texts, advance organizing statements may be presented as skeletal schemata to activate the awakening of conceptual traces of existing schemata in cognitive structure. These advance organizers may be written in the form of superordinate context or topic sentences. Context provides additional information for relating concepts to each other, and it may be utilized to improve meaningful verbal learning. Schemata within memory may be awakened through the use of advance organizers containing superordinate context. This form of context usually contains more general, abstract, and inclusive information than other sentences which are relational, as in the case of coordinate context. Superordinate context produces a facilitating effect for learning and retention as meaningful material is introduced before subsequent material. Possible organizing effects may be anticipated when written passages containing this information are presented. Importantly, the utilization of superordinate context may be significant for the introduction of subsuming material in initial learning (Gagné and Wiegand, 1970).

Advance Organizers

Another important aspect of meaningful learning involves context. In the form of an advance organizer, context provides a cognitive scaffolding for the anchorage of new concepts (vis. Ausubel and Fitzgerald, 1961). This is particularly helpful when generally related concepts are not present in cognitive structure and the material to be learned is unfamiliar. Gagné and Wiegand's (1970) study of meaningful verbal learning and hierarchical levels of context provided an insight into the use of different levels of context statements. The utilization of topic sentences in the form of superordinate context involving more general, abstract, and inclusive information facilitated the introduction of subsuming material in initial learning. Indeed, topic sentences helped learning more than relational sentences (coordinate context), because superordinate context improved learning and retention when meaningful information was introduced in the form of advance organizers before subsequent information. However, the advance introduction of superordinate context does not always significantly affect learning and retention. Other key factors affecting learning and retention may be content difficulty and student unfamiliarity with concepts to be learned (Ausubel and Fitzgerald, 1961). Another area of concern could relate to the location of organizing statements in instructional sequences and the identification of appropriate schemata for optimal anchorage of relevant existing concepts.

Gagné and Wiegand's (1970) investigation also pointed to the possible benefit of utilizing a context cue with topic sentences presented as superordinate context. This cue consisted of an extra instruction, such as "This sentence tells what the next few are all about." This particular statement preceded topic sentences in initial learning and in the retrieval of information from memory during testing. As a retrieval cue, it led to the recall of meaningful information which had been learned. Topic sentences can facilitate learning and retention when meaningful material is introduced ahead of other information, and advance organizers are used to arrange material within a program.

Advance organizers may be expository or comparative. When the learner is exposed to unfamiliar material, an expository organizer will provide an explanation. In the case of relatively familiar material, a comparative organizer may be used to discriminate between new and existing concepts that are different but confusable so that material to be presented is strengthened through an improved internal organization. Organizers may also be presented as interspersed questions. Rickards (1975-76, p. 617) also points to the importance of using superordinate context when questions are inserted in a passage of text adjacent to related subordinate material.

Context statements must be written with a clear understanding of what the user already knows about the subject to be described. The instructional designer's responsibility in this instance is to determine what is already known and then prepare advance organizers that will awaken more inclusive concepts in memory as computer dialogue is presented.

Relational Concepts and Skeletal Schemata

Explanations of the interrelationship of meaningful verbal and pictorial stimuli are not easily derived since a variety of instructional concepts are present in learning. Relational concepts may call for deductive verbal explanations that would not be needed in learning simpler concepts (Carroll, 1964). A learning differential may also be present in meaningful verbal and pictorial learning: Studies of verbal and pictorial learning show that the presentation of words and pictures together facilitates the learning of words more than pictorial recognition (Hartman, 1961; Jenkins, Neale, and Deno, 1967; Kale and Grosslight, 1955; Paivio and Yarmey, 1966). The results of these investigations suggest that verbal and pictorial stimuli, which are presented simultaneously in meaningful material to be learned, may facilitate meaningful verbal learning more than pictorial learning.

Relational verbal ideas are often found in computer dialogue, and coordinate context may be used for the purpose of making these ideas easier to understand and for clarifying subsequent ones. Organizing verbal statements provide the learner with context information about the subject to be understood. Therefore, an appropriate area of interest for software/courseware authors, that should receive some emphasis, is the design and selection of skeletal schemata that will form a conceptual framework and awaken cognitive schemata.

Skeletal schemata require careful construction. They must match the structure of the content to be learned with a pre-existing cognitive structure. The relational context of the elements in a sequence of concepts to be learned must be designed to build more fully upon the ideational scaffolding that will hold these propositions.

Conclusion

Meaningfulness of communication in computer-assisted instruction is an important area for computer education research. The influence of advance organizers in pictorial learning, their optimal placement, and the effects of written organizing statements relating to subsequent meaningful verbal and pictorial information and the identification of appropriate cognitive schemata are all important topics of concern for research, and they are of possible benefit to the development of instructional materials. Thus the relationship of meaningful verbal and pictorial learning and retention needs to be investigated more fully. Moreover, computer graphics has assumed a popular role in supporting or providing messages for computer communication. There are abundant opportunities for new research in verbal and pictorial meaning and the study of key variables as they relate to an individual student working at a computer terminal. The meaningfulness of verbal and pictorial stimuli is a paramount concern for the instructional design and development of computer software/courseware.

Computer technology allows instructional designers, programmers, and authors to interact with the reader/user of text in ways qualitatively and quantitatively different from text used more traditionally in other school learning situations. All authors, of course, write with the unseen audience in mind and, in the case of educational materials, must aim for a "common ground." The computer, if not releasing the author from this restriction, loosens the constraints considerably. The challenge, however, is greater: The author must anticipate a "range" of competencies with respect to the topic and the design of sequences for accommodating the ideational structures of each. Designers may to some degree be technicians, but the effective interweaving of meaning into syntactical structures and instructional strategies requires that they also be master designers of instruction. □

References

Anderson, R.C. The Notion of Schemata and the Educational Enterprise. In R.C. Anderson, R.J. Spiro, and W.E. Montague (Eds.), *Schooling and the Acquisition of Knowledge*. Hillsdale, NJ: Erlbaum, 1977, 415-431.

Anderson, R.C., Spiro, R.J., and Montague, W.E. (Eds.) *Schooling and the Acquisition of Knowledge*. Hillsdale, NJ: Erlbaum, 1977, 415-431.

Ausubel, D.P. *Educational Psychology: A Cognitive View*. New York: Holt, Rinehart , and Winston, 1968.

Ausubel, D.P., and Fitzgerald, D. The Role of Discriminability in Meaningful Verbal Learning and Retention.

Journal of Educational Psychology, 1961, *52*(5), 266-274.

Carroll, J.B. Words, Meanings, and Concepts. *Harvard Educational Review*, 1964, *34*(2), 178-262.

Franzwa, D. Influence of Meaningfulness, Picture Detail, and Presentation Mode on Visual Retention. *A V Communication Review*, 1973, *21*(2), 209-223.

Gagné, R.M., and Wiegand, V.K. Effects of a Superordinate Context on Learning and Retention of Facts. *Journal of Educational Psychology*, 1970, *61*(5), 406-409.

Glynn, S.M., and Britton, B.K. Supporting Readers' Comprehension Through Effective Text Design. *Educational Technology*, 1984, *24*(10), 40-43.

Hartman, F.R. Recognition Learning under Multiple Channel Presentation and Testing Conditions. *A V Communication Review*, 1961, *9*(1), 24-43.

Jenkins, J.R., Neale, D.C., and Deno, S.L. Differential Memory for Picture and Word Stimuli. *Journal of Educational Psychology*, 1967, *58*(5), 303-307.

Kale, S.V., and Grosslight, J.H. Exploratory Studies in the Use of Pictures and Sound for Teaching Foreign Language Vocabulary. Technical Report SDC 269-7-53, U.S. Naval Special Devices Center, Port Washington, NY, 1955.

Koen, F. Verbal and Non-Verbal Mediators in Recognition Memory for Complex Visual Stimuli. Center for Research on Language and Learning Behavior, University of Michigan, 1969. ERIC: ED 029 349.

Lindsay, P., and Norman, D. *Human Information Processing* (2nd ed.). New York: Academic Press, 1977.

MacDonald, N.H., Fraser, L.T. Gingrich, P.S., and Keenan, S.A. The Writer's Workbench: Computer Aids for Text Analysis. *Educational Psychologist*, 1982, *17*, 172-179.

Neisser, U. *Cognition Reality: Principles and Implications of Cognitive Psychology*. San Francisco: Freeman, 1976.

Paivio, A., and Yarmey, A.D. Pictures Versus Words as Stimuli and Responses in Paired-Associate Learning. *Psychonomic Science*, 1966, *5*(6), 235-236.

Piaget, J. Genesis and Structure in the Psychology of Intelligence. In D. Elkin (Ed.), *Six Psychological Studies*. New York: Random House, 1967.

Piaget, J., and Inhelder, B. *The Psychology of the Child*. New York: Basic Books, 1969.

Rickards, J.P. Processing Effects of Interspersed Advance Organizers in Text. *Reading Research Quarterly*, 1975-76, *11*(4), 599-622.

Whorf, B.L. Languages and Logic. In J.B. Carroll (Ed.), *Language, Thought, and Reality: Selected Writings of Benjamin Lee Whorf*. Cambridge, MA: The Technology Press of Massachusetts Institute of Technology; and New York: John Wiley & Sons, Inc., 1956.

Winn, W. The Meaningful Organization of Content: Research and Design Strategies. *Educational Technology*, 1981, *21*, 7-11.

Designing Interactive, Responsive Instruction: A Set of Procedures

Maria Harper-Marinick and Vernon S. Gerlach

In every instructional system two subsystems operate, an instructional or teaching subsystem and a learning subsystem. The relationship between the two subsystems aims at one goal: the learner's attainment of the objectives. It is well known that humans learn by doing. The teaching-learning relationship, consequently, must be based on active responses from both learners and instructors.

Active responding, while a necessary condition, is not a sufficient condition for optimum instruction. The responding must be interactive. Interaction takes place when the learner does something in response to the teacher as well as when the teacher does something in response to the learner. For instance, if one wants to teach learners to use a rule, the algorithm might look like this:

1. Teacher displays the rule.
2. Teacher presents a problem.
3. Teacher asks for the answer.
4. Learner answers.
5. Teacher tells whether or not the answer is correct.
6. Teacher asks for the rule.
7. Learner states a rule.
8. Teacher displays the rule.
9. Teacher presents a problem.
10. Teacher asks for the answer.
11. Learner answers.
12. Teacher tells whether or not the answer is correct.
13. Etc.

This algorithm can result in the development of interactive instruction. For instance, 4 will not appear until 3 has been displayed; the teacher (3) does something with which the learner (4) interacts. Or 5 will not appear until 4 has appeared; the learner (4) does something with which the teacher (5) interacts. However, this instruction, although it is interactive, is not necessarily responsive as written here; it does not take into consider-

Maria Harper-Marinick is Research Assistant, and Vernon S. Gerlach is Professor of Instructional Psychology, College of Education, Arizona State University, Tempe, Arizona.

ation the nature of the learner's response. Whether the learner answers correctly or incorrectly in 4, the teacher does exactly the same thing in 5. No matter how competent the learners, all receive the same problem in 9.

Responsive instruction is the remedy for this deficiency. The sequence of instruction needs to be determined by what the learner does and not by a preordained path specified by the instructor or the instructional designer.

The simplest form of responsiveness is feedback. Research shows that good instruction requires the application of theories on feedback. The knowledge of the correctness or incorrectness of a response following instruction increases learners' performance (Anderson, Kulhavy, and Andre, 1972). More complex forms of responsive instruction may involve feedback and some form of branching: skipping ahead, repeating a task, doing remedial practice.

Instruction that uses data about the learner as a basis for branching optimizes the amount of time the learner spends to achieve mastery (efficiency) and increases the probability of avoiding gaps and irrelevancies in the stimulus materials needed to produce accurate performances (effectiveness) (Schutz, Baker, and Gerlach, 1964).

The algorithm for teaching a rule, presented above, will produce, at best, interactive instruction. To add the potential for responsiveness, we might modify it as shown in Figure 1.

This modified algorithm illustrates a prescription for instruction that is both interactive and responsive. It is interactive because the learner responds to something the teacher does, and the teacher responds to what the learner does. It is responsive because the teacher's responses depend on the nature of the learner's actions. For instance, the direction of instruction takes a different path depending on the learner's response to question 4. If the answer is correct, the learner is presented with a new and different problem; if the answer is incorrect, the learner is provided with remedial instruction.

A Set of Procedures

In this article, we describe a five-step set of procedures for designing and developing interactive, responsive instruction. The five steps are Objective, Content, Questions, Boundaries, and Entry Behavior (Gerlach and Cooper, 1986). Because the first (Objective) and last (Entry Behavior) steps have been well described by many different authors, and because the procedures discussed in this article are not specific to any particular method of stating objectives or entry behaviors, those two topics are not elaborated. However, they are

Interactive Video

39

Figure 1

Algorithm for Teaching a Rule, Using Interactive and Responsive Instruction

1. Teacher displays the rule
2. Teacher presents a problem ◄──────────────────────────────
3. Teacher asks for the answer
4. Learner answers
5. If learner answers . . .

correctly	incorrectly
6. teacher acknowledges correct response	6. teacher provides correct response
7. teacher asks for the rule	7. teacher presents a problem
8. learner states the rule	8. learner answers correctly . . .
9. teacher presents a problem	9. go to . . .
10. learner solves problem	

incorrectly . . .
9. teacher provides answer and displays rule
10. teacher asks for the rule
11. learner repeats the rule
12. go to . . .

essential steps, and failure to attend adequately to them during the design process will generally produce unsatisfactory instruction.

Content

Once the objective and entry behaviors are determined, the designer must specify the content. The content is the *information that the learner must process in order to master the skills or behavior stated in the objective.* The content comes either from the designer or from a subject matter expert.

The content may take different forms. It may be a definition, a description, a generalization, a principle, or a rule. For instance, for the objective "divide a proper fraction by another proper fraction, given a problem in the form $a/b \div c/d = ?$, without reducing the quotient," the content would simply be the rule "invert the divisor and multiply."

A definition of "declarative sentences" and "interrogative sentences" would be the content for the objective "identify declarative and interrogative sentences."

Questions

Questions are *stimuli that elicit responses from the learner.* These questions can be stated in either interrogative or imperative form. Examples of interrogative questions might be, what is the divisor?, is this a telling sentence?, what do you do first?, should this word begin with a capital or a small letter? Imperative questions might include

such examples as mark the words that should be capitalized, underline the declarative sentences, find the quotient.

The questions perform two functions: (1) they enable the learner to interact with the content, and (2) they serve as a test for the ultimate attainment of the objective.

The first type, en route questions, enables the learner and the teacher to interact. This interaction is nothing more then the processing of the content, described above, by the learner. For instance, en route questions for the content "invert the divisor and multiply" would include "what is the divisor?," "invert it," "change $\div$ to X," "to solve $2/3 \div 1/2 =$, what do you do first?"

Mastery questions, the second type, enable the learner to demonstrate attainment of the objective without any extraneous or facilitating cues. For instance, mastery of the objective dealing with division of a proper fraction by another proper fraction would be demonstrated by a correct response to the question "solve the following problems: $2/3 \div 4/5$, $5/8 \div 7/10$."

Questions serve as a guide to what the teacher needs to ask the learners in order to help them progress toward mastering the task described in the objective.

When designing instruction for a specific objective, an initial pool of questions might be developed. An easy way is to "work backwards" "... from the target objectives to analyze the sequences of skills to be learned" (Gagne and Briggs, 1979, p. 150) and ask "What must a person do to show attainment of objective X?" Any other method for determining hierarchy or structure might be used too. Development of the question pool is not specific to any method of task analysis. Depending on the learner's performance, this initial pool of questions might be revised; questions might be added or deleted as needed.

The information obtained by using the questions from the pool is the basis for a prescription for the next step for the learner and/or instructor, which might include repetition of a task, remedial work, introduction of new material, or additional activities for enrichment.

Boundaries

However, further constraints are needed in order to make the questions as effective as possible. The designer should avoid both gaps and irrelevancies in the instruction. The question pool, therefore, must be defined by a set of boundaries. The boundaries *define the area covered by the objectives*.

To illustrate, let us use an objective which deals with teaching the identification of equilateral, isosceles, and scalene triangles. The boundaries must specify, among other things, the sizes, positions, colors, and other characteristics of the triangles displayed as part of the questions. These boundaries deal with the *domain* (or stimuli) of the questions.

The boundaries must also address the responses of the learners. The entire set of learner responses that satisfy the requirements of the performance or product specified in the objective constitutes the *range*. The instructional designer needs to know the range in order to make instruction truly responsive. For instance, given the question $1/8 + 3/8 = ?$, instruction would differ depending on whether the learner's response was 4/8 or 2/4 or 1/2.

To carry the example even further, instruction would be even more efficient if it were responsive to an answer such as 4/16. While the answers 4/8 and 2/4 require only that the learner be taught something about reducing to lowest terms, the answer 4/16 requires that something about adding fractions be taught.

The problem of specifying boundaries is by far the most technical step in the design rules, although not necessarily the most difficult. A detailed treatment of the procedures for specifying boundaries is presented in Gerlach and Cooper (1986).

_____

Conclusion

"If education is to meet the needs of individual students through provision of appropriate knowledge and training in important skills, there must be increased dependence upon well designed, effective instruction." (Dick and Carey, 1978, p. 4)

The purpose of *well designed* instruction is to ensure the existence of the conditions necessary for the development of the individual's full potential in his/her unique way. Instruction should provide learners with the opportunities to develop their capabilities to perform intellectually, socially, and physically in their environment. Learners must be participants rather than spectators in this process.

Interaction and responsiveness are two necessary conditions of well designed instruction. Interactive instruction allows learners to participate *actively* in the process, and responsiveness ensures that learners' needs are met.

We have described a five-step set of procedures for designing and developing instruction—Objectives, Content, Questions, Boundaries, and Entry Behavior. The implementation of this set of procedures will result in specifications for *interactive* and *responsive* instruction. □

_____

References

Anderson, R.C., Kulhavy, R.W., and Andre, T. Conditions Under Which Feedback Facilitates Learning from Programmed Lessons. *Journal of Educational Psychology*, 1972, *63*, 186-188.

Dick, W., and Carey, L. *The Systematic Design of Instruction*. Glenview, IL: Scott, Foresman and Company, 1978.

Gagne, R.M., and Briggs, L.J. *Principles of Instructional Design*. New York: Holt, Rinehart, and Winston, 1979.

Gerlach, V.S., and Cooper, M. *Procedures for the Development of Computer Instruction Specifications*. Syracuse: ERIC Clearinghouse on Information Resources, 1986.

Schutz, R.E., Baker, R.L., and Gerlach, V.S. *Measurement Procedures in Programmed Instruction*. (Final report NDEA Title VII Project 909.) Washington, D.C.: USDHEW, NDEA Title VII, 1964.

Interactivity in Microcomputer-Based Instruction: Its Essential Components and How It Can Be Enhanced

Herman G. Weller

Recent developments in microprocessor technology have led to the production of several types of interactive teaching tools: microcomputer assisted instruction and simulations, interactive videotape, and interactive videodisc. The use of these electronic tutoring tools is steadily becoming more widespread in education and in industrial training. These tutoring tools allow the learner to interact with the tutor, in effect, to have a dialogue with the television monitor in which a combination of computer-based instruction, computer animation, video still frames, and/or video motion images present the lesson.

There has been, however, an important drawback to the rapid development of these interactive technologies, at least to the instructional designer: instructional designs have been too often driven by the hardware technology. Educational designs and techniques today are frequently only reactions to the developing technology. The technology has emerged, and instructional developers have endeavored merely to apply the technology for instructional purposes.

In contrast, principles of the systematic design of instruction suggest that the designers must clarify their instructional requirements and goals before considering the organization, strategy, and delivery mechanisms for meeting those requirements and goals. Jonassen (1985) has proposed that "rather than creating problems to which we can apply our most popular interactive technology, we need to develop design processes which identify the required components of interactive, adaptive instruction."

Herman G. Weller is a doctoral student in the Division of Curriculum and Instruction, Virginia Polytechnic Institute and State University, Blacksburg, Virginia.

Indeed, in the hands of expert instructional developers, interactive computer-based hardware has been demonstrated to be a very powerful instructional technology. Many researchers of computer-assisted instruction (CAI) have established empirically the finding that there is an improvement in learning efficiency that occurs relative to standard instructional methods (Ebner, Donaher, Mahoney, Lippert, and Balson, 1984). In the videodisc area, the "available evidence suggests that the interactive videodisc is a highly effective instructional medium across all types of educational and training applications" (Kearsley and Frost, 1985).

Unfortunately, there is a great diversity in what the producers, customers, and consumers of these new electronic tutoring technologies interpret as *interactivity*. The term interactivity is being used, and misused, widely nowadays as a justification for the selling, buying, and use of microcomputer-based instructional tools. Too often, what has been placed on disks and videodiscs as instruction has actually been less interactive than the print-based programmed instruction courses that were developed in the 1950's and 1960's.

What Is Interactivity?

In general, interactivity enables learners to adjust the instruction to conform to their needs and capabilities. The learner becomes an active participant, rather than a passive observer, making significant decisions and encountering their consequences. More specifically, interactive lessons are those in which the "learner actively or overtly responds to information presented by the technology, which in turn adapts to the learner, a process more commonly referred to as feedback" (Jonassen, 1985).

Cohen (1984) has expanded on the burdens that interactivity places both on the interactive program and on the learner. Her definition of interactivity requires that the learner make some sort of qualitative response in order for the instruction to continue, that the instruction be dependent upon the learner's entry and be "designed to accommodate many different styles of learning, many different types of responses, and many different pathways through the program."

Interactivity Encourages Active Learning

Learning is an active process. Learners necessarily must process information actively in order to comprehend and remember it (Ausubel, 1960). Therefore, the more mentally active the learners are as they process information from the computer-based program (i.e., as they interact with the materials they are trying to comprehend), the more likely they are to comprehend them.

Kearsley and Frost (1985) list several factors which help produce a high level of student participation and involvement. These include the kind of user-control provided, the relevance of the instruction, and the instructional strategy used. The more control the learners feel they have, the more involved they will be in the instruction.

Components of Interactivity

Perhaps the most obvious aspect of interactivity is the *quantity* of interactions that occur per unit time during a program. However, more important than the rate of interactions is the *quality* of the interactions. The fact that interactivity makes the instructional sequence dependent on the learner's responses implies that *branching is the crux of interactivity*.

The quality of the interactions depends on many factors, perhaps the most important of which is the degree of learner control. Interactivity is enhanced by greater learner control of both pacing (time) and sequencing (branching). However, merely to provide learner control by means of repeated standard statements is to make the program boring. A variety of types of responses required from the learners keeps them in an active state of learning.

The instructional software program should be designed with specific learning outcomes in mind. The instructional events of the lesson, with which the learners must interact, should be based on a model of events of instruction that are related to known learning processes. The learners should be provided with an overview of the lesson as an advance organizer for their thinking. They should be given the option of fast or slow learner paths, rather than being forced to follow a linear path throughout the lesson.

The evaluation of the learner's responses and the consequent feedback from the program are important contributors to the interactivity of the lesson. The program should be easy to operate and free of programming errors which can result in its "crashing" for incorrect learner responses. The readability of the screens can affect the student responses, with well-designed screens serving as a motivating factor for the learner. The effects on learning of each of these components of interactivity will be discussed below.

Quality and Quantity of Interactivity

The rationale behind the development of the print-based programmed instruction was that causing a student to become actively involved with the learning situation increases the probability that he or she will learn. The Programmed Instruction Movement lost momentum partly because it required an extensive amount of systematic and empirical development to be effective, and partly because the programs were often boring and the learning machines broke down frequently. Similarly, the future of microcomputer-based instruction will depend upon the quality of the software developed for it.

One measure of how a software program engages the learner in the instructional sequence so that he or she is actively involved with the program is the number of interactions occurring per unit time. Although the quantity of learner responses is a significant indicator of interaction, if most of those interactions are merely motor responses and not tied in with the learning, then the interaction rate loses its importance.

Bork (1982) has stressed that the *quality* of the interaction in a program can be assessed in terms of the type of input required of the learner during the interaction, the method of analyzing the response, and the action taken by the program after the input. Bork maintains that the quality of interaction in the design of the program is the important determinant of the quality of the instructional program produced. Thus, the effectiveness of the microcomputer-based instructional program is a function of the quantity *and* the quality of the interactivity, techniques which must function together to ensure comprehensible instruction.

Interactivity Is Based on Branching

An important feature of all types of computer-based materials is that the learner may select a nonlinear path through the content according to individual abilities and needs. The student can interact with the material by choosing a sequence and direction through the instructional program. The quality of the interaction is enhanced if the content is organized so that the learner has the option of switching repeatedly between instructional sequences.

From the software-design point of view, all forms of interactivity are based upon branching (Kearsley and Frost, 1985). Branching occurs through the use of two different types of interaction: multiple-choice response or free response. For either case, branching to another part of the program occurs when the user has selected an option.

The customary multiple-choice convention is provided by either a menu or a multiple-choice question. Very often this type of condition is developed in a rather unimaginative fashion. Some innovative non-print manners of selection in menu conditions include the use of peripheral devices such as a joystick or mouse, (or a mannequin, as in Hon's (1983) CPR learning system), or the use of a touch screen.

A multiple-choice exercise is indicated when the question involves a discrimination task or a very long response. However, an open-ended question requiring free-form input by the learner corresponds more closely to the real world, and should be employed whenever possible. Accepting free-form input necessitates a more elaborate programming procedure for recognizing a correct answer, for example, accepting properly positioned commas, decimal points, and plus signs in a numerical answer.

The art and science of software design requires a delicate balance between a proliferation of unnecessary branching sequences that are rarely used by learners, and a paucity of branches to satisfy learner needs. The most efficient design procedure is probably to institute branches only for student mistakes that seem most likely, then to add branches during the formative evaluation when unanticipated student difficulties arise. Also the designer should avoid the repetition of stock feedback phrases, and select specific messages randomly from a collection of equally appropriate phrases.

Learner Control Can Reduce Anxiety

The computer or the software can control the sequencing and pacing of an instructional lesson. However, when the learner is given control over these two aspects, it can serve as a motivating factor by decreasing anxiety and improving attitude (Steinberg, 1984). It can transform a computer-based learning situation that might become frustrating into an interesting and challenging event.

The optimal degree of learner control is determined by learner characteristics, the nature of the content, and the complexity of the learning task (Hazen, 1985). Whereas some content areas require a high degree of sequencing, others do not. For example, when less able or less experienced learners are engaged in the memorization of information, more direction from the software may be necessary.

Learner Control of Display Time

Control by the learner of display time is much more desirable than computer control of it. Timed responses or screen changes often lead to learner frustration, for example, when the student has not finished reading a passage that has just vanished, or wishes to continue to the next screen while the computer is still pausing. To allow learner control of pacing and/or screen erase, a message similar to "Press any key to continue" can prompt the learner to proceed.

Learner Control of Sequencing: Pausing and Menu Access

The learner should usually be allowed to page either forward or backward through the material within a menu item. For example, this could be provided with a page-up key, page-down key, or function keys. The learner should usually also be given rapid access to the menu for the current section, and to a glossary or help screen if one is available. This could be provided for via another function key.

Learner Control of Sequencing: Instructional Events

It is not enough that the instructional events of the program be designed with the learner in mind; the learner should be provided adequate information concerning the events that he or she can best select and correlate (Hazen, 1985). At the outset, the program should specify its purpose and goals, so the student can understand clearly what will be covered. An overview of the entire lesson should be presented, including the possible paths through the lesson. This could be done by proceeding through menu items.

Not only should the organization of the lesson be well-developed, it should be presented clearly to the learner. This can be accomplished with advance organizers that relate the ideas of each section to concepts the student has learned previously. A program lacking these initial items of information on its organization is often confusing.

Each component of the lesson should be designed in light of specific learning objectives (Gagne and Briggs, 1979). The objectives might include, for example, expanding the learner's knowledge of the subject matter, enhancing a motor skill, or learning a concept. The behavioral objectives should be formulated in terms of the learning outcomes desired for the lesson component. Guidelines for the possible types of learning outcomes should be followed, such as the following ones provided by Gagne and Briggs: verbal information, motor skills, intellectual skills (comprising discrimination, concrete concepts, defined concepts, rules, and problem solving), attitudes, and cognitive strategies.

Once the learning outcomes for a lesson have been specified, the designer should utilize a model of events of instruction that specifies stages of learning as a guide for the selection of the stages of computer displays that enhance learning. Each of such instructional events need not necessarily be included in each lesson. However, each should be considered and, if omitted, be done consciously rather than inadvertently.

For example, Gagne, Wager, and Rojas (1981)

have proposed a system for planning and authoring lessons in computer-assisted instruction that are supported by the following nine events of instruction, with the related learning processes shown in parentheses: gaining attention (alertness), informing the learner of the lesson objective (expectancy), stimulating recall of prior learning (retrieval to working memory), presenting stimuli with distinctive features (selective perception), guiding learning (semantic encoding), retrieval and responding (eliciting performance), reinforcement (providing informative feedback), assessing performance (cueing retrieval), and enhancing retention and learning transfer (generalizing).

The most useful sequencing of computer instruction is often an elaboration from the simple facts, concepts, procedures, rules, or principles to the more detailed facts, concepts, principles, or procedures (Merrill, 1977). As new concepts are added, they should be related to the previously learned structure. When teaching concepts, the designer should utilize a concept attainment model that has been shown to be effective. For example, the following sequence is very useful (Tennyson and Park, 1980): (1) definition of the concept, stated in terms of the concept's critical or essential attributes, (2) an expository presentation consisting of sets of example-nonexample pairs that range from easy to difficult and are divergent, (3) an interrogatory practice presentation consisting of "newly encountered" examples.

The quality of interaction is improved if the learner is following a lesson path that is appropriate to his or her learning speed. The program should adapt in different ways to the fast or experienced student than to the slow or inexperienced student. There are several methods by which learning can be individualized:

- An optional pretest before a lesson, to determine possible paths.
- A required posttest for each lesson, allowing bypasses of material based on the learner's performance on past lessons.
- A variable number of examples according to the learner's performance or experience.
- Alternative paths which the learner can choose for presentations at different levels of difficulty.

One or more of these methods should be employed so that the student is not compelled to follow a linear path through a lesson. The total lesson path will not only be specific to each learner, but also will conserve student learning time.

Acting on the Learner's Response

Whereas the long-range consequences of a series of the student's responses involve the provision to him or her of an appropriate pathway through the lesson, the immediate feedback for each response contributes to the "ambiance" of the learning experience. Although learner responses can often be characterized as correct, anticipated incorrect, and unanticipated incorrect (Hazen, 1985), in some cases the learner's response is only necessary as evidence of thinking on some point. The program should often allow variations in the format of a correct learner response, including slight errors in spelling, case, punctuation, and grammar.

In addition to messages simply informing the learner whether the responses are correct or incorrect, motivational messages should be interspersed to which the learner may react at an affective level (e.g., "Very good"). They should be mature, positive, and varied. Following an anticipated incorrect response, instructional feedback should be provided which is corrective or remedial, and includes explanations, hints, or cues toward the correct response. Remediation should be differentiated from the less effective technique of review, where the program merely repeats the same sequence that had been viewed initially.

In the case of unanticipated responses (e.g., the learner presses "G" when the possible responses are "A-D"), the student should be given only one or two more opportunities to enter an anticipated answer. The learner should not be put into the frustrating situation of having to repeat one item numerous times until hitting upon the correct response.

Facilitating Operation

Interaction is facilitated by a program that the student can operate easily with minimal effort spent on learning its operation. Directions for operation should be accessible to the learner whenever needed via a menu. The software should be free of programming errors and able to handle all incorrect responses without "crashing."

Readability Enhances Learner Responses

Easy-to-read, well-designed screens can serve as a motivating factor for the learner. According to Hazen (1985), the text in an easy-to-read screen should possess the following characteristics:

- Well phrased.
- Double-spaced whenever possible.
- Always in a consistent pattern of indentation and justification, centered or left.
- Structured in natural eye-movement sequences, top-to-bottom and left-to-right.
- Consistently implemented and consistently placed functional screen areas.
- Utilizes message design principles, like message clarity.

It is necessary to add that all words should be spelled correctly, sentences should be grammatically correct, and text should be punctuated properly. Screen formatting and graphics should be used for readability and educational purposes; not just for decoration or clever effects. Use graphics for a specific instructional purpose, for example to convey a concept or to focus attention on an important point. Readability is improved by using different colored words, boxes, bars, or windows for emphasis and/or organization. Instructions should be separated on a screen from the main text area with windows, in order to cue the learner to always look at the same portion of the screen for the same types of information.

What Lies in the Future?

The various microcomputer-based instruction media are in their developmental stages. Designers of materials for these instruction media should strive to utilize their full potentials. An important aspect of investigation is the effect of the quality of the interaction between a learner and the program on the effectiveness of the instruction.

When more comprehensive results of educational research on many of the essential components of interactivity for microcomputer-based media are available to materials designers, then the average quality of the appropriate instructional materials should improve greatly. In turn, the impact of microcomputer-based instruction on the schools should take a significant upturn. This may bring about a systematic integration of microcomputer-based instruction into the school curriculum and classroom at all levels K-16. □

References

Ausubel, D.P. The Use of Advance Organizers in the Learning and Retention of Meaningful Verbal Materials. *Journal of Educational Psychology*, 1960, *51*, 267-272.

Bork, A. Interaction in Learning. In *Proceedings of NECC '82*. Columbia, Missouri: The University of Missouri, 1982.

Cohen, V.B. Interactive Features in the Design of Videodisc Materials. *Educational Technology*, 1984, *24*(1), 16-20.

Ebner, D.G., Danaher, B.G., Mahoney, J.V., Lippert, H.T., and Balson, P.M. Current Issues in Interactive Videodisc and Computer-Based Instruction. *Instructional Innovator*, March 1984, 24-29.

Gagne, R.M., and Briggs, L.J. *Principles of Instructional Design* (2nd ed.). New York: Holt, Rinehart, and Winston, 1979.

Gagne, R.M., Wager, W., and Rojas, A. Planning and Authoring Computer-Assisted Instruction Lessons. *Educational Technology*, September 1981, *21*(9), 17-26.

Hazen, M. Instructional Software Design Principles. *Educational Technology*, November 1985, *25*(11), 18-23.

Hon, D. The Promise of Interactive Video. *Performance and Instruction Journal*, November 1983, 21-23.

Jonassen, D.H. Interactive Lesson Designs: A Taxonomy. *Educational Technology*, June 1985, *25*(6), 7-17.

Kearsley, G.P., and Frost, J. Design Factors for Successful Videodisc-Based Instruction. *Educational Technology*, 1985, *25*(3), 7-13.

Merrill, M.D. Content Analysis Via Concept Elaboration Theory. *Journal of Instructional Development*, 1977, *1*(1), 10-13.

Steinberg, E.R. *Teaching Computers to Teach*. Hillsdale, NJ: Erlbaum, 1984.

Tennyson, R.D., and Park, O. The Teaching of Concepts: A Review of Instructional Design Research Literature. *Review of Educational Research*, 1980, *50*(1), 55-70.

Part Three

Design and Production
of Interactive Video

Design and Production of Interactive Videodisc Programming

Eric A. Davidove

Introduction

The ideal instructional development process calls for the selection of media based, in part, on the ability of the selected media to present instructional events. Most instructional development models advocate the practice of first gathering data, then defining the problem, developing solutions, and evaluating and modifying solutions as needed. Most models are based on the premise that the design should determine the media. However, there are times when the media must determine the design. For example, when educational institutions procured personal computers for instructional purposes, a need was expressed to design more computer-assisted instruction. The selection of the medium (e.g., personal computers) superceded the identification and definition of the desired learning outcomes. Thus, the instructional events were planned to fit the parameters or capabilities of the personal computer. If media are selected at the onset, prior to a front-end analysis, designers must adopt a different or revised development process.

In August of 1984, the Instructional Development and Evaluation Center (IDEC) at Gallaudet College set out to explore applications of a prototype level III interactive videodisc system. The delivery system (e.g., interactive videodisc) was selected prior to the identification and definition of a topic or application. The development processes illustrated in almost all instructional design models were not appropriate for the project assigned to IDEC. A modified or revised approach was created. This article describes a 14-step development process which was constructed on the basis of insight gained later. The purpose of this revised development process is to assist future developers involved with projects in which the decision to use interactive videodisc-based instruction supercedes the identification and definition of desired learning outcomes.

The recommended design process described in this article represents only one of many possible approaches for solving an instructional problem. The model came about as a result of completing one interactive videodisc design project and has not yet been tested for effectiveness.

Eric A. Davidove is an Instructional Developer/Designer, Instructional Development and Evaluation Center, Gallaudet College, Washington, D.C.

Steps of the Development Process

STEP 1: Learn the Authoring Capabilities.
STEP 2: Select a Topic/Application.
STEP 3: Prepare Project Implementation.
STEP 4: Conduct a Front-End Analysis.
STEP 5: Design Application.
STEP 6: Test Program Design.
STEP 7: Develop Scripts for Video and/or Audio, and Computer Graphics.
STEP 8: Preproduction Formative Evaluation (Level 1).
STEP 9: Produce Video Master (Motion and Still Frame Sequences).
STEP 10: Produce Computer Graphics.
STEP 11: Merge and Test—Formative Evaluation (Level II).
STEP 12: Master and Replicate.
STEP 13: Summative Evaluation.
STEP 14: Implement and Maintain.

STEP 1: Learn the Authoring Capabilities

The design team members should become familiar with the media element treatments available on an interactive videodisc system and with advantages and disadvantages inherent in particular treatments. The analysis of advantages and disadvantages of media element treatments during the first step is based solely on technical considerations. That is, what can and cannot be accomplished with the selected interactive videodisc system? Step 5 calls for an analysis based on instructional design factors such as the desired learning outcomes, audience profile, and learning styles.

The most common types of media elements available on interactive videodisc systems can be grouped into three categories: (1) input devices, (2) output devices, and (3) an information management.

Input Devices. Input devices are media elements which enable a user (learner) to respond or enter information. This includes elements such as a touch-sensitive screen, keyboard, mouse, light pen, or voice recognition.

Output Devices. Output devices are those media elements which present visual and audible stimuli to a user. This category of media elements includes visuals from a videodisc, computer graphic, and/or printer, and sound from a videodisc, compact disc, or voice chip. These stimuli are presented to the viewer via output devices which can operate alone or in trandem with one or more of the other options.

Information Management. Information management is necessary when a user's performance or interaction with the system needs to be identified, collected, stored, and analyzed. The media elements in this category are indicators or switch settings and log files. Switch settings, or gatekeepers, are used to branch a user according to previous responses or bookmark entry and exit points, and log files document the user's pathway, performance, and use of input devices.

STEP 2: Select a Topic/Application

There are probably as many topics as there are ways to select them. This section describes possible criteria measured during the topic and application selection process. The degree to which the criteria are prioritized will vary.

Audience Profile. The audience profile includes variables such as age, reading level, physical abilities, home environment, native language, gender, and educational history in light of the topic. Some attributes are more important than others. Priorities and criteria should be established and applied during the topic selection process.

Performance Type. The program goals and learning objectives should be defined in a broad sense at the outset. However, every attempt should be made to classify them by performance type. The performance types, or domains, reveal the necessary external conditions. The types of conditions necessary for learning a motor skill differ from that of learning a defined concept. Some external conditions are appropriate and obtainable with interactive videodisc systems, while other conditions might be more effectively provided with a book or lecture.

Needs. A need is defined as a discrepancy or gap between what is and what ought to be (Burton and Merrill, 1977; Kaufman, 1976); that is, the difference between what the learners already know how to do and what the objectives state they should be able to do. The designer attempts to identify, define, and write down these needs in terms of broad goals and priorities.

Endangerment. Conducting an experiment, operating a table saw, and working on an engine could be dangerous activities for a novice. A lesson design which has a high probability of inflicting bodily injury on the student should be presented in a foolproof manner. Interactive videodisc systems could provide a simulated experience to novice students prior to a dangerous hands-on application.

Budget Constraints. Expenditures could be minimized if alternative delivery systems are explored. It is not always necessary to purchase materials or hire professional actors. For example, a biology student studying the anatomy of a frog does not necessarily have to dissect a real frog. A simulated lab environment could be provided with interactive videodisc technology. Conversely, expensive chemicals and equipment do not have to be purchased if the same experiment can be successfully presented on an interactive videodisc system, and actors or guest speakers do not have to be employed every time a demonstration of a scene or concept is provided.

Time Schedules. Most course offerings are presented during finite time-lines. The amount of time a student attends a course or training session is predetermined and usually inflexible. This could be a problem if the process of learning a skill or concept does not fit into the preset time schedule. For example, a pre-med student cannot wait for long-term effects of a prescribed treatment, and a biology student cannot afford to wait weeks for a chemical reaction to take place. The consequences of decisions which normally take too long for practical time constraints must be condensed. This can easily be accomplished with interactive videodisc technology. Real-time situations can be compressed and presented within scheduling constraints.

Assumed Learning Environment. The assumed learning environment is the location where the instruction takes place. This might be a classroom, an outdoor site, a warehouse, or a learning carrel. Some features to look for in a learning environment, in light of the learning objectives, number of students, and necessary materials, are: (1) distance to the location, (2) size of the location, (3) number of electrical outlets, (4) quality of lighting, (5) security, (6) location of the materials, and (7) available support staff.

Assumed Development Environment. Identification of an optimal location for the design team members to work while developing instruction is made by considering: (1) number of members, (2) available support services, (3) budget and time constraints, (4) size of the location, (5) electrical requirements, and (6) necessary equipment and materials.

Related Existing Materials. When working under budget, time and resource constraints, an available supply of existing related materials becomes a priority. The design team should make an effort to identify a pool of available material which can be easily incorporated into interactive videodisc-based instruction. This includes materials such as charts, slides, video and film footage, and scripts.

STEP 3: Prepare Project Implementation

The goal of the project manager during this phase of the design process is to establish a roll-out plan which stipulates each team member's discrete responsibility and time schedule. First, a formally organized meeting with all project participants is arranged. During this meeting the project manager reviews major program objectives, explains each phase of the design and development process, sets a project schedule, and clarifies all "next steps."

STEP 4: Conduct a Front-End Analysis

During step 2, the design team members identified program goals and learning objectives, as well as other criteria to assist them during the topic selection process. Step 4 calls for a more in-depth analysis; a front-end analysis. The purpose of this analysis, as it is used in this design process, is to continue to identify and define what to teach. A good way of approaching this task is to draft a description of all the desired learning outcomes. The question that designers usually ask of content specialists is "What will the learners be able to do after the instruction that they could not do prior to the instruction?"

Information Process Analysis. The purpose of this analysis is to reveal the sequence of mental operations in performance of the desired learning outcomes; terminal or target objectives (Gagné, 1977). Terminal objectives are performances to be reached only by the end of the course.

Task Classification. Learning outcomes (objectives) are categorized into domains in order to identify the conditions of learning. The five major types of domains are motor skill, attitude, verbal information, intellectual skill, and cognitive strategy (Gagné and Briggs, 1979). The external conditions necessary for learning differs with each type of domain.

Learning Task Analysis. The goal is to identify the enabling objectives for which instructional sequence decisions need to be made (Gagné and Briggs, 1979). Enabling objectives are smaller in scope and are required in order to attain a target objective.

A final phase in this step of the design process is to develop performance measures that check pupil progress. The intention is to evaluate the effects of the instruction in terms of the stated performance objectives.

STEP 5: Design Application

Decisions are made about content flow, necessary external conditions, interactive category, and evaluation guidelines. At the end of this step, the design team should know *what* is taught, the sequence of teaching, and *how* the teaching occurs. This information is reflected on a flowchart.

Flowchart. The flowchart is used as a guide and communicative tool among the designer, the content specialist, the video director, the script writer, the data entry person, and the artist. A flowchart requires the assembly of information that is not only legible and helpful to the members of the design team, but also both educationally and technically sound. The flowchart is the hub of the development of interactive videodisc-based instruction. All other design tasks connect to it, and product development cannot proceed without it.

Interactive Category
1. Menu Driven: used for reference or catalogue applications, point of sale or marketing programs, or point of information exhibits.
2. Test Driven: employed with programmed instruction and assessment based programs.
3. Exercise Driven: applicable for simulations, problem-solving situations, and tutorials.

Evaluation Guidelines.
1. Record-Keeping/Tracking: devise a method of tracking and documenting a user's performance and/or interaction with the instruction.
2. Scoring and Remediation: periodically score the user's performance and provide appropriate branching (remediation).
3. Gatekeeper Questions: identify the important questions where mastery is necessary, keep track of the user's performance, and branch him or her accordingly.
4. Certification Levels: design instruction in accordance to certification specifications (medical students, law students, and so on).

STEP 6: Test Program Design

By this point in the development process, it is essential to conduct a test run of the program design. The purpose of a test run is to check the accuracy and smoothness of branching and transitions, to detect errors with the way in which media elements are entered during the authoring process, and to provide members of the design team with a better "picture" of the program. A more formal pedagogical review and analysis takes place after script development (video, audio, and graphics). The next step is a brief discussion about each team members' role during the test run process.

Author. The author makes certain that every possible branch activates according to the specifications depicted on the flowchart. The author should not be concerned with the instructional validity or quality of the branching, (i.e., whether the learner receives the appropriate type of review or feedback). The purpose is to review the program to make sure all media elements occur where and when they are planned. Finally, the author notes distracting or rough transitions and makes suggestions or revisions.

Designer and Content Specialist. Designers and content specialists check to see if the external conditions seem to be adequate. They check timing, content, feedback, variety, and numerous other features that influence the learning process.

Production Staff. Artists and television support people study this early program test in order to acquire a better feel as to the flow. They specifically focus on transition routines, framing possibilities, color, timing, and other factors that effect delivery of a message and composition of a picture.

At this point, there is no videodisc, and a majority of the graphics are still in paper form. For purposes of this test any videodisc could be used. The important thing is to have video appear and remain on the screen as it would in the final program. As far as graphics are concerned, the test requires only "quick and dirty graphics," which simply state a summary, name, or code referenced on the flowchart. The artist might want to experiment with various formats to get a getter idea of the time, placement, and size constraints. This is especially true with overlays, touch screens, and keyboard input.

STEP 7: Develop Scripts for Video and/or Audio, and Computer Graphics

Video and/or Audio Production. The television director, designer, and artistic director are the key decision-makers during the development of scripts for video and/or audio production. These decisions set the precedent for the remainder of the design process. The next section describes a few factors that merit consideration by members of the design team.

Set Description: Initial set design suggestions should be made by the artistic director and the television director. The artistic director is responsible for maintaining a color signature throughout the program. The television director is responsible for maintaining the budget, time, and manageability levels within design constraints (i.e., storage, set-up, break-down, lighting, materials, shipping delays, production staff). The designer and content specialist provide feedback about the relationship of the message and the desired learning outcome. For example, a picture, bright colors, or distracting movements behind a speaker are not appropriate for a program to teach receptive sign language skills to the hearing impaired.

Costume Design: The same procedure for set design is applicable to costume design. Costumes should be "real" to the situation, within design constraints (i.e., budget, time, manageability), and augmentative to the color signature.

Duration of Scenes and/or Still Frames: The television director should review the test run of the program and refer to the flowchart. It is relatively easy to enter estimates of running times into the time line and get a feeling of the program flow. Some factors which influence decisions regarding the duration of video sources are the available space on a disc, voice-over, sound effects, music, the audience profile (attention span, learning style, etc.), importance of the message, and the objective of the segment.

Shot Size-Framing: Some factors that effect the size of framing of video sources are graphic overlays, touch screen areas, and keyboard input. Therefore, the television director and artistic director should work hand in hand while defining the size and frame parameters.

Action on Screen: The moving video sources should be storyboarded so that the beginning and ending of each segment is clearly established. The movement or screen direction should also be noted in order to maintain consistent and smooth edits.

Order of Shots/Sequence: The order of video sources may vary depending on the user's selections or responses. However, each possible branch point should be reviewed and considered during the scripting and storyboarding phase.

Audio: All audible information (voice-over, music, dialogue) should be timed and strategically placed within the program design. Audio may reside on one of the two audio tracks located on the videodisc, on a compact disc, or in a digitized format on a hard or floppy disc.

Computer Graphic Production. Several decisions about computer graphics need to be made during this step of the design process. Most of the computer graphic specifications are documented on the flowchart. This list below describes each feature of a computer graphic.

Size: Size considerations are especially important when employing graphic overlays. Precise location of windows, transparent colors, and text cannot be specified until video production is complete, but rough estimates should be made so that the video sources are framed properly.

Text: All textual content and locations (if possible) should be clarified. Typical graphics with heavy textual content include menus, test items and directions, advanced organizers, summary sections, and title screens. Text, as described here, also includes use of prompts.

Color: Establishment of the color theme (palette), and system and debug colors should be made before graphics are created. Notation as to the colors made transparent or opaque should be made.

Wipe Routine: Transitions from a graphic to another graphic or video source and vice versa are controlled by wipe routines. The way in which wipe routines change usually vary with speed. This speed is easily controlled by the author. Wipe routine possibilities are an important factor to consider while developing both graphic and video sources. Exciting, attractive, and effective juxtaposition is possible when wipe routine features and possibilities are analyzed.

Image Description: At this point, it is helpful to have closure on the type of images that appear on each graphic. This is a descriptive decision as opposed to a review of a rough draft. A decision might be that two men in their thirties sit in an office by a computer, or a middle aged oriental woman demonstrates cooking to a group of children, etc. These descriptions are what the artistic director takes back to the drawing table to produce rough drafts.

Touch Areas: Initial decisions about the number, location and size of touch areas should be reflected in early renderings of the computer graphics.

Name: Providing a name to each graphic as it is being developed will save time and make organization easier. As soon as possible, even before graphics are complete, the author should begin to enter the graphics by name. Each graphic should also be printed and catalogued.

STEP 8: Preproduction Formative Evaluation (Level I)

A formative evaluation should provide data which enables the design team to identify specific ways to revise and improve the materials, the lesson plans, the performance tests, and indeed the operation of the entire instructional system. The development process described in this article employs two levels of formative revision and improvement. Level one starts with individual evaluation; level two solicits feedback from a small group during a trial implementation.

The first level of the formative evaluation process enables detection of necessary changes prior to the production phase. This minimizes carry-over of mistakes. The second level follows immediately after individual evaluation, with emphasis not only on educational effectiveness but also on administrative feasibility (Johnson *et al.*, 1985).

The first level of the formative evaluation is commonly referred to as the clinical phase (Briggs and Wager, 1981). The video sequence should be put into storyboard format, rough sketches of computer graphics should be rendered and be made available for review, and tests (directions and questions) should be written. The designer and content specialist sit with an individual learner and together they "page through" the instruction. The learner asks questions, provides comments, points out confusing or difficult sections, etc. Whenever possible, the designer and content specialist fill in the instructional gaps and note them. As a result of this phase, gross problems are identified and noted for further analysis and correction. This process reveals and detects student learning difficulties, gaps in the material, and necessary prerequisite knowledge.

It is not the purpose of a formative evaluation to compare the developing instruction with a finished product. On the contrary, the purpose is to identify ways to improve the instruction *as it is being developed* (Briggs, 1977).

STEP 9: Produce Video Master (Motion and Still Frame Sequences)

The production of the video master can be viewed in three major phases. These phases and their associated tasks are discussed in the following sections:

Pre-Production Planning. The production of a video master requires considerable preparation. The following tasks should be completed, but the order in which they are started (or finished) varies with each project.

Location Identification: All indoor, studio, and outdoor locations should be identified and secured. It is advisable to have a back-up set for all outdoor locations in the event of bad weather, and all outdoor locations should be scheduled for shooting first.

Set and Costume Construction: All sets, props, and costumes should be constructed and incorporated into rehearsals as early as possible. The actors should work with their "tools," the director should work out blocking, and the artistic director should assess the appearance of shapes and colors on camera.

Talent Selection: Audition notices should be sent out. Studio time should be booked for taped auditions. Scripts should be prepared. Call backs and final selections should be made. All actors should sign an agreement stating that they will be available for any re-shoots and accept payment equal to that of the regular shoots. This enables the design team to revise scripts after evaluating a check videodisc.

Disc Geography: Video source location on the

disc affects the program flow. Every effort should be made to lessen the random access time. The disc geography should be reflected in the shoot sheets, whenever possible, in order to expedite editing time.

Shot Sheet Development: A shot sheet is a document that lists the order in which video sources are produced (video and audio). The development of a shot sheet is affected by the weather, location, actors' availability, and the disc geography. For example, all outdoor shoots are scheduled first, high-paid "talent" are in and out quickly, etc.

Teleprompter Script Development: This might be necessary for both on and off camera talent (i.e., voice-over). Before making a Teleprompter script, it is important to see which type face, size, and location on the paper projects clearest on the Teleprompter. Actors should see and use the scripts prior to production.

Character Generator: All screens originating from a character generator should be made, stored, and catalogued in preparation of the transfer to video tape.

Rehearsal Schedule: The television director, television staff, actors, artist, designer, and content specialist should be present at rehearsals. Each member of the design team should look for flaws based upon previously established criteria. This is a good time to work out a lot of the "bugs" so as to avoid wasted time and money during all phases of the production. Some portions of the program should be rehearsed outside a television studio (i.e., classroom, office) since studio time is expensive.

Production/On-location Shoot. Mistakes in content, omission of a scene, and technical problems should be caught before the actors are dismissed and the set is torn down. During the taping sessions, a designer and content specialist are responsible for identifying problem areas. These responsibilities are discussed below.

Script Supervision: The content specialist needs to make sure the dialogue coincides with the script. Actors occasionally deviate from the script, and qualitative judgments of content accuracy need to be made.

Duration: The designer should keep a tally of individual times and total program times. This is especially true in instances where all 28 minutes of disc space is needed.

Premaster Specifications. The designer should check the premaster videotape to see that the audio levels, color levels, and other technical factors outlined in a document issued by the premaster facility, are in accordance to the proper standards.

Post-Production. The outcome of post-production is an edited master videotape and a check videodisc. The following section lists and describes the step by step process.

Log Dailies: The dailies are the raw footage before editing occurs. Each tape must be logged. A typical log sheet calls for the SMPTE time code numbers (start and end), the tape source or number, a short description of the scene, still frame, audio track contents, and notation of any technical difficulties.

Review of Dailies: The content specialist, designer, and television director should select the "good takes." A second log sheet, which coincides with the planned disc geography, should contain a comprehensive listing of all the acceptable video and audio sources. This means that the video and audio sources occupying the outer tracks on the videodisc are listed first on the log sheet.

Review Final Edit. As each team member reviews the final edit, they should be looking for technical errors, content problems, omission of video or audio sources, and improper placement of images and audio.

Produce Check Videodisc. A back-up copy of the final edited videotape version should be made and the original should be sent to a facility that makes check videodiscs. The direct read as written (DRAW) process seems to be the least expensive route.

Log Check Videodisc. Most videodisc players have the capability to read frame numbers. Each video source (video and audio) should be logged in a similar fashion as the dailies.

STEP 10: Produce Computer Graphics

There is a wide variety of graphic software packages on the market. Criteria for selection of the "best" package can be derived by conducting an analysis on the type of graphics needed, the competency or background of the artist, budget, and time frame.

Competency or Background of the Artist. The selection of a graphics package is affected by the competency or background of the artist. Some artists have never used a computer before, some prefer to use a stylus, some prefer to work through an assistant.

Budget and Time Constraints. There are basically three types of software packages. One is a digitizer, a second is a graphics tablet or mouse-driven package, and a third is a keyboard-driven package. Each has its advantages and disadvantages.

The *digitizer* is useful when a quick, realistic rendering of a complex item is needed (i.e., circuit board, engine part). However, it is more expensive and complex than other types of packages.

The *graphic tablet and mouse-driven* packages are great when free-hand renderings are needed. They are usually accepted by an artist with no computer background, and they offer many possibilities. In addition, new styles of fonts are easily created by either working from pre-established designs or starting from scratch. The disadvantages are that the stylus and tablets can break. The broken items may take weeks to repair, and costs are usually high. Another disadvantage is that only one person can work at a time, assuming one tablet or mouse is purchased.

The *keyboard-driven* packages are the most practical but also the most limiting in some respects. These packages are advantageous when working with text, adding color, and designing geometric shapes. More than one person can use the package at a time, assuming there are enough computers. The costs are relatively low. However, some shapes and patterns are virtually impossible to make with keyboard driven packages.

STEP 11: Merge and Test—Formative Evaluation (Level II)

As mentioned earlier, level two of the formative evaluation process is conducted with a small group. Prior to the tryout, the check videodisc and graphics are in a semi-polished form and integrated into the system. Sometimes it is helpful to solicit review of the video sequences while still in the videotape format. Other times it is best to proceed with a review using a check videodisc. The cost of making a check videodisc is relatively low (usually about $300.00), and the time delays which occur while searching on a videotape format might be too long for an effective assessment.

Integrate Computer-Based Elements with the Check Videodisc. Enter the remaining elements into the program design. This includes the frame numbers on the videodisc, the location of graphic windows, keyboard input windows, the location and size of touch areas, and the definition of transparent colors. All other media elements have been entered during Step 6.

Run the Presentation. The presentation (program) should be reviewed to identify authoring errors. This includes the timing and speed of video/audio sequences, transitions, the location of windows, touch areas, graphic overlays, keyboard input spaces, voice-chip responses, the type and speed of wipe routines, the accuracy of branching, and so on. The first analysis should be strictly technical in nature. Verification of authoring is assessed in light of the flowchart and other pertinent documentation.

Conduct Small-Group Tryouts. Small-group tryouts range from five to twenty learners. It is not preferable to have a self-selected group. The designer and content specialist should establish criteria and select (or hire) students with attributes similar to that of the target audience. Another factor that effects the selection process is the students' availability (time schedule).

All instruction should occur in the same manner as that of the assumed learning environment. A space should be set aside and proper furniture, work space, tools, support staff, etc., should be available. The environment is a major concern because one important aspect of a level two evaluation is the administrative feasibility.

Solicit Responses. The instructional designer and evaluation specialist should design measurement tools and establish methods for obtaining feedback from content specialists, members of the target audience, artists, designers, and the television director. The data-collection instruments should gather information on those areas that can be changed, revised, and improved upon (things you *can* do something about). This includes training time, preparation time, study time, testing time, test results, reactions, technical difficulties, student interaction with the videodisc system, manageability, and the work-station arrangement.

Revision. Once all the problems have been identified and possible solutions explored, the revision process can begin. Revisions might include changing the order of video sequences on the disc, an omission of a scene, a color change in graphics, the amount of time allowed per sitting, the arrangement of a work station, and numerous other possibilities. It is crucial to set aside money and time for the revision process. As mentioned earlier, it is unwise to collect formative data that cannot be applied to the revision process.

Verify Changes. Whenever possible, try and verify the changes made as a result of the formative evaluation (before producing a master videodisc). The video and audio changes can be verified in four ways: (1) reviewers (learners, production staff, content specialist, designer) view a videotaped version (when quick random access time is not crucial for the analysis); (2) reviewers watch a live presentation (a reading, acting out, or use of visuals); (3) reviewers page through the instruction; or (4) reviewers see an updated check video-disc version.

STEP 12: Master and Replicate

The turn-around process for making a master videodisc from videotape varies according to the facility used and the amount paid. One way to save time is to recheck the premaster videotape for adherence to the facility's published specifications. Most facilities are willing to work with a design team prior to pre-production so as to avoid technical difficulties and time delays during the mastering process.

STEP 13: Summative Evaluation

The purpose of a summative evaluation is to determine the value of the system. People want to know if the students learn what they are supposed to learn, if they learned it faster, if they retained it longer, and if they enjoyed it more than other students learning a different way. Instructors and administrators want to know if it is easy to manage, what the preparation and maintenance costs are, and if it is worth their time and energy. Generally the issues relate to pedagogical and economic concerns.

STEP 14: Implement and Maintain

Implementation and maintenance involve variables from selling the product to installing enough outlets in the learning environment. Managers need to be trained, schedules need to be arranged, budgets need to be set, students need to be recruited, and so on.

Conclusion

The 14-step development process described in this article might be misleading. On the surface it seems to imply that one step begins as another ends. On the contrary, during a typical development process, the designer recycles many steps in the process and constantly revises previous work. Hence, backward projections, revisions, forward projections, and more revisions must be made in light of insights gained later.

Development of instruction on an interactive videodisc system is costly, labor-intensive, and complex. However, the capabilities of the technology are promising and the resulting instruction can be exciting and effective. □

References

Briggs, L.J. (Ed.) *Instructional Design: Principles and Applications.* Englewood Cliffs, NJ: Educational Technology Publications, 1977.

Briggs, L.J., and Wager, W.W. *Handbook of Procedures for the Design of Instruction (2nd ed.).* Englewood Cliffs, NJ: Educational Technology Publications, 1981.

Burton, J.K., and Merrill, P.F. Needs Assessment: Goals, Needs, and Priorities. In L.J. Briggs (Ed.), *Instructional Design: Principles and Applications.* Englewood Cliffs, NJ: Educational Technology Publications, 1977.

Johnson, J.F., Wildwequist, K.L., Birdsell, J., and Miller, A.E. Storyboarding for Interactive Videodisc Courseware. *Educational Technology*, 1985, *25*(12), 29-35.

Kaufman, R.A. *Needs Assessment: What It Is and How To Do It.* San Diego, CA: University Associates, 1976.

Knirk, F.G., and Gustafson, K.L. *Instructional Technology: A Systematic Approach to Education.* New York: Holt, Rinehart, and Winston, 1986.

Gagné, R.M. Analysis of Objectives. In L.J. Briggs (Ed.), *Instructional Design: Principles and Applications.* Englewood Cliffs, NJ: Educational Technology Publications, 1977.

Gagné, R.J., and Briggs, L.J. *Principles of Instructional Design* (2nd ed.). New York: Holt, Rinehart, and Winston, 1979.

Project Management Guidelines to Instructional Interactive Videodisc Production

Martha R. Tarrant, Luke E. Kelly, and Jeff Walkley

Introduction

This article describes the planning process for an interactive videodisc module in motor skills assessment and, more specifically, the formulation of a documented template to guide project managers and instructional technologists through the maze of organizational tasks involved in similar productions.

Beyond presenting yet another application example, the authors' intentions are (1) to provide resources for design activities which seem so often to be taken on faith, and (2) to reiterate some of those resources which, for us, were particularly thought-provoking in relationship to our particular task. As Parkhurst and Dwyer (1983) state, "[i]t is readily held that most decisions concerning the design and structure of instructional software are made on the basis of intuition, feelings, past successes, or on some other visceral basis" (p. 10). Our hope is that by providing a start-to-finish project scheme based on literature largely from this magazine and two primary sources, we may give some grounding for instructional software products that reflect on our entire field.

Rationale for the Program

The ability to accurately assess qualitative performance of motor skills is critical to the Physical Education professional. Yet, typically, professional preparation courses have not provided the best instructional opportunities for developing that ability. Classes in assessment are often teacher-centered and conducted in large groups. As such, they may be able to convey the "rules" of per-

Martha R. Tarrant is a doctoral student in Instructional Technology and a video producer, Curry School of Education, University of Virginia, Charlottesville. **Luke E. Kelly** is assistant professor and project director, Department of Human Services, University of Virginia. **Jeff Walkley** is a lecturer in charge of Adapted Physical Education, Phillip Institute of Technology, Bundoora, Victoria, Australia.

formance and observation without the practice, self-testing, immediate personalized feedback, and remediation needed to establish competency. An individualized, competency-based module which could furnish necessary instruction in performance criteria as well as multiple, evaluated practice trials in observing and assessing performance would greatly enhance preservice Physical Education programs.

Medium and Referent Models

In preparing for the creation of an assessment module for the Adaptive Physical Education Program at the University of Virginia, it was determined prior to any other design considerations that interactive videodisc (IVD) was the medium of choice. Technically, IVD ensured access speed, search accuracy, video quality, and durability unavailable with tape. Of all the media considered, IVD also created the most realistic representation of motor skill performance and, therefore, the best simulation of actual job conditions related to assessment of that performance.

The disc's capability for quality freeze frames and frame-to-frame "stepping" enabled the interactive program to provide several presentation alternatives to real time video. Examples in real time were deemed necessary but not sufficient for learning because while assessment in the field (after instruction) would be in real time with each skill executed in no more than 1-2 seconds, the learner (during instruction) might not be able to discriminate between discrete subskills without stop-action or slow motion.

Given the selection of medium, a literature review at that time (1984) revealed only two detailed process models specific to interactive videodisc design. One—*Interactive Videodisc Design and Production Workshop Guide*—was an adaptation by WICAT, Inc., of the military's "Interservice Procedures for Instructional Systems Design" (Campbell et al., 1983). The second was the invention of Michael DeBloois and the basis for his book, *Microcomputer/Videodisc Courseware Design* (1982). As neither model was clearly superior to the other and both omitted tasks which were felt to be important to this project, the instructional designer chose to merge the best of the two with some practical additions from other sources. The result, which resembles DeBloois in form but not strictly in content, is the 80-step PERT chart shown in Figure 1.

Caveats

The authors offer three caveats with respect to Figure 1. First, as suggested above, it is not intended to be a new or "revolutionary" model. It is a hybrid of existing referent design systems and

Figure 1

Interactive Videodisc Project PERT Chart

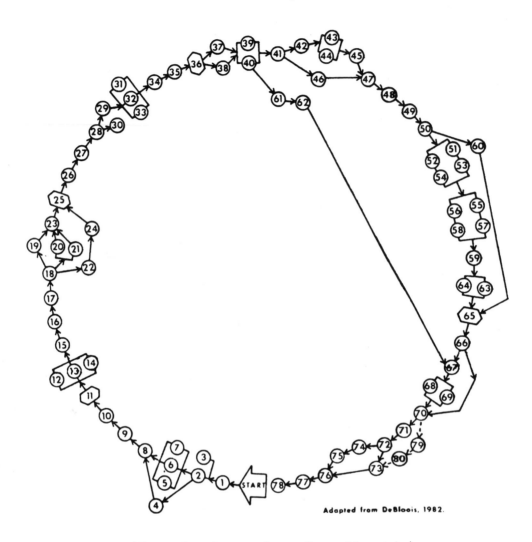

Adapted from DeBloois, 1982.

(For explanation, see Appendix to this article.)

principles. It can best be described as an "evolutionary model" (Andrews and Goodson, 1980) which, while research-based and grounded in theory in so far as its sources are grounded, is virtually untried and unvalidated as a unit apart from the project for which it was derived.

Second, it is intended to be procedural, not systemic. Each step is not necessarily dependent on the step immediately prior to it but logically grows out of the progression of prior activities. Some steps can and do overlap; some can be rearranged or even omitted to conform to the needs of the individual project. As long as the integrity of the basic activities is maintained, there is flexibility without sacrificing thoroughness.

Third, it is highly detailed because it was drafted as a "one-time-only" plan. Finite limitations imposed on the project meant there had to be a practical beginning and end to the timeline as opposed to a continuous, self-perpetuating loop of evaluation and revision.

That is not to say that the project manager or instructional designer is discouraged from re-entering the network of events to review any or all of the project plan if time allows. In fact, the steps were purposely arranged in a cyclical pattern (after DeBloois) to suggest that option. But client considerations of resources, costs, and immediacy of need may demand closure at some point, which is provided here.

Figure 2

Fault Analysis Tree

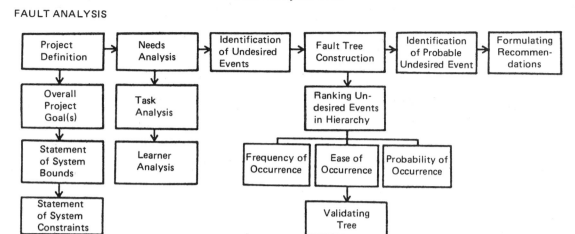

FAULT ANALYSIS

Adapted from Wood, Stephens, and Barker, 1979

Template Overview

Certain activities merit highlighting because they were either new to DeBloois and WICAT, not new but presented in greater depth, or substantially different from them in the ways in which they were applied:

Fault Analysis

The fault analysis or "failure assessment" scheme in events 11, 25, and 36 originated with Woods, Stephens, and Barker (1979), was borrowed by DeBloois, and has been modified by us for what we felt was simplicity's sake (Figure 2). One of several evaluation tools in the template, fault analysis periodically scrutinizes the global elements of the project and from them extracts all the foreseeable potential problems with regard to the plan of action, learner objectives, and media selection. Those "undesired events" are then ranked in order of severity by how frequently, easily, and likely each would be to surface if the design were left unchanged. Beginning with the highest priority problem and working top-down, the team members then have the opportunity to "brainstorm" solutions and ideally have ways to incorporate those solutions into the design to prevent the hypothesized difficulties from ever taking place.

A simple example of the benefits of fault analysis can be derived from our project's assessment of the plan of action. After funding was

secured and three graduate students had committed to act as team members along with the project manager, a tentative timeline and resource plan was drafted (event 10) with April 1 as the targeted start-up and July 31 as the completion date of the development phase. During the process of identifying undesired events, we discovered that WICAT advised a minimum 20-week, full-time allotment for an IVD module of average complexity, while we allowed only 16 weeks of part-time participation for all members of a novice team. Our first fault assessment, therefore, revealed that we may have had unrealistic time expectations. This "failure" was later justified as we experienced delays in the bid process (through which we obtained specialized hardware), unscheduled equipment breakdowns, and loss of "personnel" to graduation and relocation. In fact, the development actually took closer to a year with revisions, research, and summative evaluation still on-going two years after completion.

Learner Analysis

DeBloois suggested a learner analysis approach which, again, we have changed slightly—in this case, to provide greater user control (Figure 3). Data are collected on a representative sample of intended users to identify characteristics among them that could conceivably affect their learning [event 19]. Those characteristics are then factored into homogeneous sets to describe learner groups A, B, C, etc. (events 31-33). A variety of appro-

Figure 3

Learner Analysis Matrix

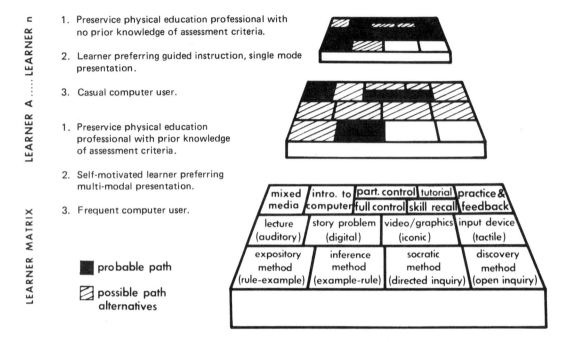

LEARNER A ...... LEARNER n

1. Preservice physical education professional with no prior knowledge of assessment criteria.

2. Learner preferring guided instruction, single mode presentation.

3. Casual computer user.

1. Preservice physical education professional with prior knowledge of assessment criteria.

2. Self-motivated learner preferring multi-modal presentation.

3. Frequent computer user.

LEARNER MATRIX

■ probable path

▨ possible path alternatives

mixed media	intro. to computer	part. control full control	tutorial skill recall	practice & feedback
lecture (auditory)	story problem (digital)	video/graphics (iconic)	input device (tactile)	
expository method (rule-example)	inference method (example-rule)	socratic method (directed inquiry)	discovery method (open inquiry)	

Adapted from DeBloois, 1982

priate media and learning style options are pooled into a matrix (event 34) and prescriptions to address individual differences for each group are then drawn from it into multi-level extensions.

If disc "real estate" were unlimited, the system would be able to accommodate multiple prescriptions per group, but more often than not, the best learning method for everyone is the only method used. DeBloois is to be credited for advocating the best method for each group, but even this is not completely satisfactory. Although the learner can control the sequence of content, movement between modes of presentation, and speed of presentation this way, he would be locked into a single subset of the program and unable to control his own strategy. He could move within his prescribed path but not out of it.

The alternatives to no learner-strategy control are full or partial control. Full control would give the user access to all group paths through the program, suggest the best one for his needs, and allow him to choose for himself. Partial control might either put limited restraints on all groups equally or allow full control to one group while limiting another. In choosing what, if any, type and degree of strategy control is appropriate, we felt the designer should add one more step to DeBloois' analysis procedure. Once the "best strategy" matrices have been formulated, an analysis of the effects for each group between strategies should be conducted to safeguard against negative effects for those who choose other than their designated path.

Computer-Managed Testing

The computer management plan (event 30), which controls the entire training system, is made up of several features, not the least of which is practice assessment and competency testing or "assessment accuracy." Figure 4 offers our management plan as an illustration of a typical schema.

Some discussion of testing is evident in both referent models but as far as we could determine, "computer-managed" testing (CMT) and feedback were not specifically addressed. Historically, testing has been a paper-and-pencil exercise and as we presumed that designing for paper-and-pencil was inherently different from designing for the electronic media, we recognized the need to review testing (event 43) and feedback (event 44) princi-

Figure 4

Computer-Videodisc Management

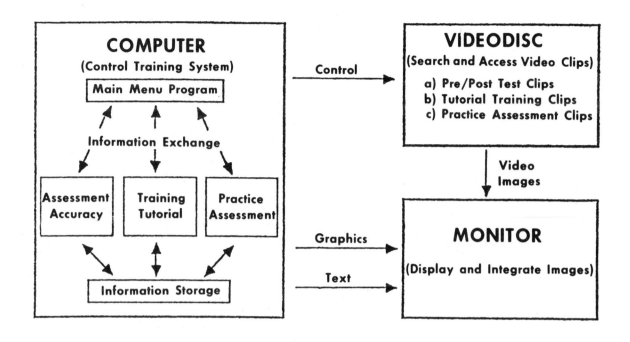

ples in light of their computer delivery. The results of that review are some guidelines on CMT/feedback extracted from Mizokawa and Hamlin (1984) and Cohen (1985):

(a) Test instructions on the computer should not exceed a single screen of 80-column characters or two sequential screens of 40-column characters with a toggle key between them (Mizokawa and Hamlin, 1984).

(b) Information about automatic timing, scoring, and help screens should be offered prior to the test, along with the practice items. When timing the test, program an audible cue at the beginning and end or immobilize the keyboard before and after the designated time allotment. If the subject completes the test before his time has expired, he should be allowed to exit by using a particular keystroke (Mizokawa and Hamlin, 1984).

(c) Test items themselves should not be divided between screens or be too complicated to work without benefit of scratch paper. The questions should be self-explanatory and the vocabulary understandable by any representative user, as individualized IVD instruction may preclude monitoring of the text by those who would otherwise provide assistance (Mizokawa and Hamlin, 1984)

(d) Immediate informational feedback should

be condensed to the least possible amount as it slows the student's progress through the test. End-of-session feedback, on the other hand, is crucial as it relates his score, identifies his mistakes, provides opportunities for remedial prescription and review while the material is current, and thus, contributes to accurate long-term retention (Cohen, 1985).

Computer Text Screens

While both WICAT and DeBloois offered general visual design ideas, we again felt that neither model was sufficient where design guidelines of computer text screens were concerned. Computer programmers, who would generate such screens, are not normally graphic artists with the creative ability to construct visually-appealing textual material without direction. For that reason, we have not only created a separate event dedicated to screen design (event 51) but also supplemented WICAT and DeBloois with design principles of other authors. The resulting guidelines are divided into three directives:

(a) Consider the type, use, and resolution of the learning environment display. Displays normally fall into the category of television receivers, monochrome or composite color monitors, and RGB monitors. In each instance, different criteria for screen image construction are recommended. With television, when the image is intended for

Figure 5

Screen Grid

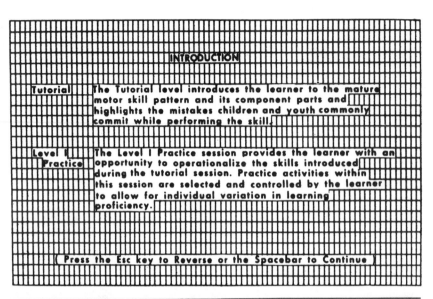

group viewing through a large screen projector, resolution will be poor. To compensate, use no more than 25 computer characters per line for 8-10 lines per "page" (Braden, 1986) or 28-32 per line for 12-13 lines from a video character generator (Campbell *et al.*, 1983). Where television is to operate at an individual workstation, the resolution—although still poor—improves with proximity to the user and could be said to be the equivalent of a medium resolution digital monitor.

Therefore, the number of characters per line could be increased to a maximum of 40-45 (Braden, 1986), although DeBloois advises no more than 35. The same set of criteria would apply to an RGB monitor that produces a sharp 80-column image when its intended use is with large-screen projection. If, however, that monitor will be viewed by a single student at a work-station, the text can be expanded to as many as 70 characters per line (Braden, 1986) filling up to 90 percent of the page's character slots (Reilly and Roach, 1986), provided that the lines are double-spaced and the letters are in upper and lower case for readability (Hathaway, 1984).

(b) Create a grid which meets the recommended image criteria for the display. Depending upon the results of this display analysis, a mock screen grid can be devised (after DeBloois' example) by marking off the appropriate number of horizontal or "character" blocks and vertical or "line" blocks on graph paper (Figure 5). Each block would, thereby, constitute a different character or spacing position. To guarantee the proper 3x4 or 4x5 proportion, an outline representing the screen borders should also be drawn. The area within should be formatted like an overhead transparency (DeBloois, 1982) with content limited to one concept per screen (Kearsley and Frost, 1985) or to a list with no more than seven qualifiers related to one concept (Braden, 1986). Lettering should be chosen for simplicity and made consistent throughout each and every screen. Variations should be limited to a small, select group of compatible graphic character fonts (Kearsley and Frost, 1985) or to the same font in different sizes for highlighting (Reilly and Roach, 1986). Be advised that a change in font may adjust the individual block size in the same way that a size change in the same font would. The length of allotted lines, width of allotted rows, and overall character-by-line configuration will be affected accordingly (Campbell *et al.*, 1983).

(c) Once the grid has been structured, apply color with felt-tip markers to copies of the grid. Students anticipate the use of color in visual displays (Campbell *et al.*, 1983). It should, therefore, be a primary component of the design when the display is either a color or RGB monitor. However, Kearsley and Frost (1985) caution that color should be used only for highlighting learner cues and then only in compatible combinations of four or less colors per screen for beginners and seven or less colors per screen for more competent users (Braden, 1986). If there is a wide disparity among users, we would advocate a simple, harmonious three-color palette with the darkest color as background, the lightest color for text (or vice versa), and the middle color as a unifying border.

Storyboard and Flowchart Pretest

Event 50, which calls for pretesting the storyboard and flowchart, is a significant addition inspired solely by author Scott Fedale (1985). Fedale proposed using a character generator (CG), videotape player, editor, and microcomputer with interface board to simulate content and flow of the

IVD program design before it goes into production. His plan calls for the designer or video producer to reproduce text and video screen displays (as nearly as possible) on the CG and then transfer those images to tape, taking note of the respective "in" and "out" cues. The computer programmer would integrate these access point notations into the flowchart and write a simple program to run the planned routines. At this point, the program would be evaluated by representative users.

Although not in Fedale's plan, we would advocate that the subject matter expert (SME) also participate in the flowchart/storyboard pretest. His input would provide a second—technical—level of formative evaluation by pointing out errors in content sequence, terminology, definitions or descriptions, message and visual interactions, activities, and results of activities. Any suggestions for revision made by him (or the user group) could then be produced in the same manner and edited into the test tape.

Events Outlined

While it is not possible within the scope of this article to describe each event in the same detail as above, the appendix suggests some underlying tasks we felt needed to be accomplished within events. The procedures represent our interpretation/elaboration of WICAT and DeBloois (as the majority of steps were in some degree part of one or the other model). However, where they were found wanting in specifics, we have provided an additional resource notation as promised.

In conclusion, any interactive videodisc project can be viewed as a business with independently operating product lines that, once developed, are assembled into a single commodity—the IVD system. Individual products (such as computer software, video, audio, graphics, and print materials) each have their own design criteria, development personnel, schedules, and quality control standards but are all judged ultimately on the merits of the end-product. To be a successful product in most industries means innovation and technical excellence. To be successful in IVD systems means meticulous planning and justification for activities as well. It is hoped that our experience of devising our own template will facilitate that process for others. □

Appendix

1. **Analyze problem.** Determine the purpose of the project by interviewing the client(s) and deciding what elements within his organization are not functioning as intended. Draft a generalized statement which justifies correcting the malfunctions.

2. **Identify tasks for instruction.** Have the client rank the malfunctioning elements by highest to lowest priority and attach a degree of proportion to each [e.g., how much more important is the top priority item than the second item on the list?] Decide which of the important problems can be alleviated with instruction.

3. **Eliminate tasks not for instruction.** Focus on the problem with the highest priority. Get client approval for instruction in that problem area. Disregard all others.

4. **Perform formal needs assessment.** Inventory problem components. Establish criteria for performance of those components by interviewing the client, SME and/or a person on the job with a qualitative performance record. Observe the job site for gaps between criteria ["what should be"] and actual performance ["what is"]. Have the SME approve your list of discrepancies. (Kaufman, 1986)

5. **Determine if funding is available.** Analyze the discrepancies by costs to correct them, i.e., costs to design a remedial program; investment costs in hardware, support software and supplies to develop the system; production, implementation, and operating costs to run the program over time. Divide the list of costs between those that are "full" (shared overhead expenses) and those that are "direct" (applicable to this project only). Establish tentative times (both to produce and implement the program), personnel requirements, and wage levels. Generate a preliminary budget. Determine if the client has resources to fund the preliminary budget with a cushion for cost overruns. (Beilby, 1979-80)

6. **Assess your institutional capability for performing the work.** Decide if you can produce the program internally given the personnel and production facilities at your direct disposal. If not, determine how much of the work must be committed to outside help (consultants, freelance artists, production houses, etc.).

7. **Identify the time and human resources required available.** Attach a reasonable time frame to each of the problem components. Using the identified personnel, determine how much time each is able to give to the project and compare with the time needed. If insufficient, extend the timeline or find additional or alternative resources. (Marlow, 1981)

8. **Determine whether performing the work is in the interest of your institution.** Compare the work to be done with your institutional goals and objectives. Decide if it is in line with the type of projects that will further the company's aims and, thereby, promote growth. Weigh the consequences to other projects and to personnel of undertaking this job even when it is suited to your institution and decide if it will be profitable enough to justify diverting your resources from other work.

9. **Decide whether to commit resources to the problem.** On the basis of your determination, continue or abandon the project. Seek the approval of your institution's management and inform the client of your choice. If continuing, also inform personnel who will be affected by the decision.

10. **Develop a systematic plan of action.** For each of the required personnel, list the general responsibilities and allotted time frames within which they must fulfill those responsibilities. Plot the times against your project sch-

edule to be certain that all duties can be performed without overbooking [giving the individual too much to do] or overlapping [giving the individual multiple things to do at once]. Generate a copy of his role in the plan for each worker. (Carey and Briggs, 1977)

11. **Assess for failure events in the plan.** Look for hidden agendas and other flaws in the timeline, budget, availability, and dependability of in-house staff as well as outside vendors, relations with the client and SME, establishment of the problem and performance criteria, etc. Review the individual plans with each worker and the overall plan with your institution's management. (See Fault Analysis in Template Overview.)

12. **Obtain commitment for necessary additional resources.** Inform outside vendors of your intention to use their services, the products you will need for them to create and the general specifications for those products, when they can anticipate starting the job and when you expect delivery of the finished work. Identify a person in authority and obtain agreement to the process and schedule from him. (Marlow, 1981)

13. **Obtain clarification and specification of the work to be done.** Ensure that the client and SME understand the plan you are about to undertake and that you understand their needs and their instructional problem. If necessary, revisit the job site and evaluate the specified problem components through another skills performance.

14. **Determine presentation mode to be used.** Choose from "page turner" (a catalog of video or text screens accessed in any order directly from a menu), "branching" (the response-driven accessing of video or text segments related to a fixed question), "generative" (randomized selection by computer of video and text screens from a pool of available segments), "linear simulation" (a step-by-step job aid with video and text accessing of the next step only on completion of the last), and "system simulation" (a realistic model of system functions in video driven by alternative input device responses to a given situation). (Campbell et al., 1983)

15. **Hire or commit development team.** Locate an "expert" in every major aspect of your project: video production/direction (to oversee video, audio, animation, character generation, editing, post-production effects, and mastering); instructional systems design (ISD) and authoring (for instructional design, flowchart, storyboard, script, evaluation and implementation); computer programming (for management plan, data entry, disc integration, debugging); and subject matter expertise (to be responsible for content specifications and technical reviews). Once located, insure that experts are available [full-time if timeline is short or dependent on strict deadlines] and willing to work for salaries you can offer. Have them sign a contractual agreement to perform the work. (Chadduck, 1976)

16. **Schedule and conduct planning conference.** Present data already accumulated to those who have not been involved from the start of the project and encourage further analysis of needs, problem components, resources, and measurement criteria from them. Explain sequence of events to follow: development of objectives, job/task analysis, learner matrix and strategies, delivery system, program format and management plan, testing and evaluation plan, implementation strategy, etc. Specify which

team members will be involved in which of the events and have them schedule those events in their work load. (Campbell *et al.*, 1983)

17. **Obtain project authorization.** Have the client sign a contract giving you exclusive rights to perform the work.

18. **Identify specific goals to achieve.** List unique or unambiguous problems to be addressed (1) by the project and (2) by the instructional program separately—convert each problem on each list to a succinct, behaviorally-stated, and observable/measurable statement of intent. (Dick and Carey, 1978)

19. **Identify general learner characteristics.** (See Learner Analysis in Template Overview)

20. **Obtain data on in-place learning environments.** If a fixed site has been authorized for implementation of your system, visit and analyze it for criteria which will influence its design and delivery. Given the probable system size and physical characteristics, the number of potential users, and the time horizon, look for features like "(1) distance to the location, (2) size of the location, (3) number of electrical outlets, (4) quality of lighting, (5) security, (6) location of the materials, and (7) available support staff." (Davidove, 1986)

21. **Analyze tasks for most suitable environment and arrange for placement.** If there is no fixed site, use the same criteria to develop an ideal hypothesized site for your purposes. Determine if a similar site exists and if it is available. Identify costs involved in renovating the existing site or in creating a new site to your specifications. Decide if the budget and timeline can support a new site or the planned revisions to an old site. Determine if the project warrants the expense to both. Contact support services to bid on the contract. Where staffing is concerned, decide if the budget and timeline can also support retraining current support personnel or hiring new or additional qualified staff.

22. **Identify external evaluation criteria.** Locate an independent subject matter expert. Determine his content and measurement expectations for the finished IVD product after studying the problem statement and instructional goals. Draft a preliminary program evaluation instrument from his input.

23. **Develop specific project objectives.** Break down each identified goal into a precise, performance-based "target" outcome statement by which the effectiveness of the project and the instructional program can be judged. Review the project objectives with the client to make certain that you and he have the same intentions and evaluation criteria. Review the program objectives with the team SME. (Gagne, 1977)

24. **Establish internal evaluation criteria.** Compare the target objectives reviewed by key personnel within the client organization to the expectations of the independent SME. Extract and analyze any that are substantially different to determine if they should be included on the evaluation instrument. Arrange for one-on-one and small group formative evaluations of the entry behaviors, preliminary flowchart/storyboard, and master tape. (Dick and Carey, 1978)

25. **Assess project objectives for failure cases.** Look for flaws in the target outcomes and the criteria for measurement. Make certain that all objectives are clear, exact and observable, with stated performers and conditions of performance. (See Fault Analysis in Template Overview)

26. **Identify available media alternatives.** Look for existing media resources that relate to your program. Preview video, audio and graphics for pertinence, accuracy, quality, availability of master, necessary changes, copyright. (Campbell *et al.*, 1983)

27. **Perform task analysis.** Break down target objectives into subordinate subskills. Classify subskills by domain: verbal information, intellectual skill (with subclassifications of discrimination, concrete concept, defined concept, rule and problem-solving), cognitive strategy, motor skill or attitude. Establish the instructional criteria for that domain. (Gagne and Briggs, 1979)

28. **Develop hierarchies.** Decide if each instructional goal represents a procedural, hierarchical, or combined sequence of operations. Depending on your decision and working backwards from the goal, arrange its target objectives linearly right-to-left, top-down or both. Fill in the outline with subordinate subskills. Based on general learner characteristics, identify probable subskills required of the user before he can start instruction. (Dick and Carey, 1978)

29. **Specify learning events and activities.** Write an enabling objective for each subskill (including entry behaviors) using the appropriate verb for its domain. For all the domains identified in the subskills above entry level, generate tentative strategies for presentation and a list of corresponding IVD sample activities to draw upon as options for the learner matrix. (Campbell *et al.*, 1983)

30. **Develop computer management plan.** Outline the functions required of the computer program: information dispersal and gathering, access to and control of disc, testing, scoring and feedback, record storage and tracking, branching to remediation, etc. (See Computer-Managed Testing in Template Overview)

31.-34. **Learner analysis.** (See Learner Analysis in Template Overview)

35. **Make macro-media selections and match design assumptions to media capabilities.** Determine which media provide the best opportunities for presenting the pooled learner strategies. Decide which, if any, of the previewed existing media could be incorporated into your disc in view of those media-strategy matches and make arrangements for obtaining the original(s). (Merrill and Bunderson, 1979)

36. **Assess for probable failure events in media selection.** When there is more than one appropriate medium per strategy, be sure your choice is based on its ability to complement that strategy and not its convenience (availability, ease, or cost of production) or its inherent allure. Assuming that new is better than modified old material, look for errors in your review of available existing media (particularly where relevance and copyright are concerned). (See Fault Analysis in Template Overview.)

37. **Make hardware/software decisions.** Taking into account the anticipated length and complexity of the program and the selected presentation mode, choose between small, mid-level, or large system options (differentiated by scale of computer, memory capacity, type of videodisc player, number and type of simultaneous input devices, and coding technique). Decide whether to produce the software on the system for which it is designed to be used or on another system using Linear Predictive Coding

(LPC) for cross-system compatibility. Make a choice between development of the program as a single system or a dual system with one program stored on a mainframe in Programmable Read-Only Memory (PROM) for speed and security. (Campbell *et al.*, 1983)

38. **Decide upon treatment, style, learner cue format.** Make micro-media decisions. Given detailed learner analysis, adjust the skill level which each user should have attained before instruction begins. Working now from left to right (if the sequence of events is procedural) or from bottom to top (if the sequence is hierarchical), develop the pooled instructional strategies and learning activities scheme by establishing events of instruction for each skill above the revised entry level: preinstructional activities, information presentation, student participation, testing and follow-through activities. Designate video, audio, graphics, text or a combination for each component of the events of instruction. (Dick and Carey, 1978)

39. **Determine extent of local production opportunities.** List the micro-media designations separately by type. Decide how much of each type can be produced in-house and if the committed external resources are capable of producing the remainder. If not, find new or additional outside vendors.

40. **Establish requirements for commercially produced media.** Contact outside sources and request their specifications for camera-ready art, shooting script, editing sheets, etc. Inform the team member responsible for that phase of production of the specifications.

41. **Perform detailed analysis of learning structure for each learner group.** Research the components of your learner profiles for groups A, B, C, etc., and their respective relationships to your pooled strategy, sample activities, and media selections. Eliminate any strategy, activity, or medium which might adversely effect the learning process within a given group, according to your research.

42. **Finalize and draw learner hierarchy for each learner group.** Use the same research criteria to analyze the sequence of events leading to each instructional goal for groups A, B, C, etc. Eliminate any steps in the sequence which might adversely effect the learning process within a given group. Develop individualized hierarchies by pinpointing the remaining enabling and target objectives. Similarly, develop learning prescriptions for each group by pinpointing the remaining strategies, activities, and media which correlate most directly with those objectives and the learner profile. Draft a multi-level extension to the pooled learner matrix for each prescription. (See Learner Analysis in Template Overview.)

43. **Develop test items to match objectives and locate point of examination within each learning hierarchy.** Generate test items which meet the conditions of the objective, use the same verb or elicit the same behavior as that specified by the verb of the objective and demand the same performance criteria as the objective. For each unique subskill, corresponding enabling objective and assumed entry behavior in the learner group hierarchies, generate several such items of comparable content and difficulty level. (Dick and Carey, 1978)

44. **Develop feedback plan.** (See Computer-Managed Testing in Template Overview)

45. **Pretest sample of users to insure that entry be-** haviors match level of training analysis. Develop a criterion-based test instrument using only the items developed for assumed entry behaviors. If the assumed behaviors differ from group to group, develop multiple instruments. Administer the appropriate test to a prearranged small group of representative users conforming to each learner profile. Depending upon the results, keep the entry levels intact or adjust them for higher-than-expected or lower-than-expected performances.

46. **Obtain detailed content reports from the SME.** Give a copy of the learner hierarchies to the SME as an outline from which he must develop the topic of your program. Insure that that development proceeds from the simplest to most complex concepts (i.e., left to right if the sequencing is procedural or bottom to top if it is hierarchical).

47. **Develop preliminary flowchart.** Using universal computer programming symbols, graphically depict all the main messages of the content reports and connect related messages in a branching network. Add symbols related to the instruction (at points where stops for student participation, media access or concept conclusion, reviews, testing, feedback and remediation will occur) as well as symbols for computer functions (log on/off and security procedures, help screens, and management criteria).

48. **Develop preliminary script.** Have the designer/author analyze the flowchart and content reports for continuity of messages and objectives, proper sequencing, accurate branching, smooth transitions and presentation style. Write lines for each of the main messages based on the reading and comprehension level of users.

49. **Develop preliminary storyboard.** Given the flowchart, determine the rate and amount of material presented by every route. Compare these to the complexity of the instruction and the learner profiles to determine if stops to break down the flow of information or changes in the media to alternate the psychological processing mode should occur more or less frequently. Generate a description and rough draft of visuals and corresponding audio for every main message not meant to be presented by a text screen. Make a branching notation at the bottom of each drawing not in linear sequence to indicate screens before and/or after it.

50. **Pretest and revise flowchart/storyboard.** (See Storyboard and Flowchart Pretest in Template Overview)

51. **Design computer screens.** (See Computer Text Screens in Template Overview)

52. **Design still visuals.** If producing graphic art, design the visuals in 3x4 aspect ratio with no frame lines. Restrict designs to the simplest, clearest illustration with the least number of text characters possible. To overlay text without a character generator, use black letters on a white background or vice-versa. Keep the same aspect ratio as the drawing. Avoid using material from books or journals unless they have been reproduced as camera-ready art. If producing slides, design only for a horizontal format and center the important aspect of the shot as slides have a 2x3 aspect ratio and the left and right margins will drop out of the television "safe" area. Design slides with an eye to maintaining uniform brightness between sequential shots as flicker will result if there is too much variation. (Campbell *et al.*, 1983)

53. **Design motion visual sequences.** If you have the capability to use motion picture film or video, plan to produce all except problems in the motor skills domain with video as it is less expensive than film. Freeze-frames of a movement in progress are technically superior in film. As with slides, avoid designing for differences in lighting in edited or non-linear sequences.

54. **Write final audio script, shooting script.** Determine whether all audio segments interact with the appropriate visuals and whether the limits of real time audio on an interactive disc have or have not been exceeded. If the narration runs over 27-30 minutes depending upon your system, condense the script or arrange to have the audio digitized. Extract all segments of the storyboard visuals that are to be shot on video or film. Separate those segments into groups according to location and generate a split-page shooting script for each with video and production directions on the left, audio on the right.

55. **Produce still visuals.** Convert to tape if necessary. Send specifications and copies of the rough draft visuals to the appropriate in-house employee(s) or external source. If your organization or the external production house lacks the capability to transfer still visuals to 3/4" or 1" videotape, arrange to have the conversion done elsewhere. Have the designer oversee the graphic artists, photographers.

56. **Produce motion visual sequences.** Send specifications and the shooting script to the appropriate in-house employee(s) or external source. Have the team producer/director oversee the field and studio operations.

57. **Produce audio track sequences.** For audio not produced with concurrent video or film (i.e., music, sound effects, voiceovers, etc), send specifications and copies of the script and storyboard to the appropriate in-house employee(s) or external source. Have the producer/director oversee the sound engineer.

58. **Produce necessary adjunct materials.** If implementation calls for special tools such as customized input devices or support documentation, send specifications to the appropriate in-house employee(s) or external source.

59. **Obtain SME approval.** Based on his participation in the flowchart/storyboard pretest, have the SME endorse the program content and presentation as revised.

60. **Program computer courseware.** Review the hardware/software specifications, flowchart, computer functions, and computer-generated text screen designs with the team programmer. Have him generate algorithms, procedures to accomplish the algorithms, and text screens specified by the courseware map.

61. **Initiate contact with the videodisc mastering house.** Schedule disc processing according to the turnaround time estimate of the mastering facility you intend to use, providing a cushion for revisions to the master tape should it be necessary.

62. **Obtain disc production specifics.** Have the mastering house send a list of technical requirements with reference to the tape size, coding, field specifications, audio tracks and digital dump space, master tape layout, etc. Review those specifications prior to editing.

63. **Schedule editing time slots.** Arrange for off-line video editing on the basis of the number of edits planned divided by an approximate rate of 10 sequential edits per hour. Ensure that the editing suite has the major equipment necessary for you to meet the mastering house specifications: 3-1" VCRs, computer-controlled editor, broadcast quality character-generator, time base corrector, etc. (Campbell *et al.*, 1983)

64. **Edit all segments onto master tape.** Develop a plan for the master tape layout, taking into account that (1) discs are read from inside to outside, (2) the lowest resolution is at the extreme inside rim, (3) the slowest, least accurate access is at the extreme outside rim and (4) the area between should have related material as close together as possible to cut down search time and lag between branched segments. Draft an editing log with ordered segments specified by source tape number, time code in-cue and out-cue, verbal description, and segment length. Total all segment lengths to recheck that they don't exceed 30-minutes. Have SMPTE time encoded on the tape as you make the necessary transfers. Review the edited master to be certain that all segments on the flowchart and editing log were included on the tape. Draft a second log with in-cues and out-cues from the edited master.

65. **Pilot master tape under computer control with representative learning groups.** Using the master log, have the team programmer integrate video access points into his rough draft courseware program. With the computer-videotape player interface used to pretest the flowchart/storyboard, review the results with him and decide how to refine the rough transitions. Make the necessary changes and run the program for small-group formative evaluations. Survey the learner groups who evaluated the program for their attitudes toward it and suggestions for improvement. (See Storyboard and Flowchart Pretest in Template Overview)

66. **Revise and complete master tape.** Analyze and prioritize the suggestions. Determine if any high priority suggestions require new production for or re-editing of the videotape. Compare the financial and timeline costs for the revision with its perceived importance to the program and decide whether or not to revise. Arrange for revision.

67. **Send master tape to videodisc production house.** Complete the production order form and copyright infringement waiver. Authorize a purchase order or check to cover payment for the check disc. Copy the master tape and keep the copy on file. Insure the original and send it by the fastest, most reliable carrier in your area.

68. **Obtain check disc.** Have the production house press a wax release copy of the disc before mastering.

69. **Verify quality of check disc.** View the disc to confirm that all segments of the master tape were included and arranged according to the layout plan. Look for technical flaws in the original production materials which may not have been evident on the master tape (such as jump cuts, flicker, key-stoning, color shift, etc.) as well as flaws in the pressing (field-to-field incompatibility, audio-visual incompatibility, still frame jitter, etc.). If the problem is in the master, decide whether to make revisions and repress the check disc. If the problem is in the pressing, inform the disc mastering facility and request repressing at their expense.

70. **Integrate computer program with check disc and make access point notations on flow chart.** Complete the computer program by adding the final frame numbers from the disc and refining any rough areas. Test the ad-

dressability of frames. If the frames cannot be accessed accurately, determine if the inaccuracy is consistent across all frames. Have the team programmer adjust the final frame numbers in the program to conform to the slippage. If the inaccuracy is not consistent, check the program, hardware, SMPTE coding and disc for flaws. Have the flaw corrected by the programmer, vendor, editor or mastering facility.

71. **Decide whether additional pilot testing is necessary prior to authorizing disc replication.** Determine if the timeline allows for further testing when no significant changes have been made since pilot testing the computer-controlled master tape. If there have been changes which might effect the learning outcomes, inform the client of delays and run the program for a new, parallel small-group of representative users. Collect their suggestions and analyze for revisions. Use the same criteria (from # 66-71) to decide whether or not to revise.

72. **Proof system and assess revisions, if any.** Make revisions/corrections. Run the program with the hardware for which it was intended and check compatibility.

73. **Arrange for field testing of system.** Contact the client and request a sample of no less than 30 representative users who parallel the learning groups to evaluate the final program with the in-place system.

74. **Authorize disc mastering and replication.** Inform the mastering house that the disc meets your requirements. Complete the production order forms, indicating how many copies of the disc will be required. Forward a purchase order or check for the work.

75. **Obtain disc from mastering house.**

76. **Complete integration of computer program with disc.** Have the programmer validate his check disc frame numbers with the master. Run a test on the system using all possible branches. If using multiple discs and workstations, install all discs and ideally, run a test on each.

77. **Initiate field test and evaluation.** If designating one work station per user, have the 30 sample users evaluate the program together or at their convenience. If there are limited work stations for the number of testees, schedule times to avoid overlap and contamination of the results.

78. **Verify digital program and transfer to master tape** (optional). If the computer program is to be stored on the videodisc and read by an internal microprocessor in the disc player rather than by an independent computer, allow space on the master tape for digital "dumps."

80. **Create proof disc** (optional). Arrange for the mastering house to generate a second check disc with stored digital dumps.

References

Andrews, D.H., and Goodson, L.A. A Comparative Analysis of Models of Instructional Design. *Journal of Instructional Development*, 1980, *3*(4), 2-14.

Beilby, A. Determining Instructional Costs Through Functional Cost Analysis. *Journal of Instructional Development*, 1979-80, *3*(2), 29-34.

Braden, R.A. Visuals for Interactive Video: Images for a New Technology (with some guidelines). *Educational Technology*, 1986, *26*(5), 18-23.

Campbell, J.O., Tuttle, D.M., and Gibbons, A.S. *Interactive Videodisc Design and Production Workshop Guide* (Report No. ARI-RM-83-52). Orem, Utah, WICAT, Inc., 1983. (ERIC Document Reproduction Service No. ED 244 580.)

Carey, J., and Briggs, L. Teams as Designers. In L.J. Briggs (Ed.), *Instructional Design: Principles and Applications*. Englewood Cliffs, NJ: Educational Technology Publications, 1977.

Chadduck, P.H. Selection and Development of the Training Staff. In R. Craig (Ed.), *Training and Development Handbook*. New York: McGraw-Hill, 1976.

Cohen, V.B. A Reexamination of Feedback in Computer-Based Instruction: Implications for Instructional Design. *Educational Technology*, 1985, *25*(1), 33-37.

Davidove, E. Design and Production of Interactive Videodisc Programming. *Educational Technology*, 1986, *26*(8), 7-14.

DeBloois, M.L. *Videodisc/Microcomputer Courseware Design.* Englewood Cliffs, NJ: Educational Technology Publications, 1982.

Dick, W., and Carey, L. *The Systematic Design of Instruction.* Glenview, Illinois: Scott, Foresman and Company, 1978.

Fedale, S.V. A Videotape Template for Pretesting the Design of an Interactive Video Program. *Educational Technology*, 1985, *25*(8), 30-31.

Gagne, R.M. Analysis of Objectives. In L.J. Briggs (Ed.), *Instructional Design: Principles and Applications*. Englewood Cliffs, NJ: Educational Technology Publications, 1977.

Gagne, R.M., and Briggs, L.J. *Principles of Instructional Design* (2nd ed.). New York: Holt, Rinehart, and Winston, 1979.

Kaufman, R. Obtaining Functional Results: Relating Needs Assessment, Needs Analysis, and Objectives. *Educational Technology*, 1986, *26*(1), 24-26.

Kearsley, G.P., and Frost, J. Design Factors for Successful Videodisc-Based Instruction. *Educational Technology*, 1985, *25*(3), 7-13.

Marlow, E. Using External Resources. In *Managing the Corporate Media Center*. White Plains, NY: Knowledge Industry Publications, 1981.

Merril, P.F., and Bunderson, C.V. Guidelines for Employing Graphics in a Videodisc Training Delivery System. *ISD for Videodisc Training Systems First Annual Report*, 3. Orem, UT: Learning Designs Laboratories, WICAT, Inc., 1979.

Mizokawa, D.T., and Hamlin, M.D. Guidelines for Computer-Managed Testing. *Educational Technology*, 1984, *24*(12), 12-17.

Parkhurst, P.E., and Dwyer, F.M. An Experimental-Assessment of Students' IQ Level and Their Ability to Profit from Visualized Instruction. *Journal of Instructional Psychology*, 1983, *10*(1), 9-20.

Reilly, S.S., and Roach, J.W. Designing Human/Computer Interfaces: A Comparison of Human Factors and Graphic Arts Principles. *Educational Technology*, 1986, *26*(1), 36-39.

Woods, R.K., Stephens, V.G., and Barker, B.O. Fault Tree Analysis: An Emerging Methodology for Instructional Science. *Instructional Science*, 1979, *8*.

Designing a Visual Factors-Based Screen Display Interface: The New Role of the Graphic Technologist

Tony Faiola and Michael L. DeBloois

Introduction

It's 1988. Most of us who are not old enough to remember '56 Chevys, Ed Sullivan, and perhaps IBM's Coursewriter have survived the western world's transition from an industrial to an information age. The computer has permanently altered our environment, impacting it daily by information-providing technology. We may have survived the Information Revolution, but have we taken an orientation that promotes a healthy, satisfying interaction between ourselves and the technology that characterizes this new age? What strategies and guidelines have we adopted to receive, assimilate, and communicate information? Today the communicator is challenged with design tasks requiring the use of a variety of powerful and sophisticated electronic tools to attract and hold the attention of the viewer and to present complex systems of information.

This article addresses a new role for the graphic designer as a graphic technologist (GT) who specializes in electronic-based visual communications. As such, the GT's responsibility is to use visual communication principles and visual factors research findings when developing computer screen displays for interactive video disc systems (IVD). In addition, the results of cognitive and perceptual research in the areas of legibility, chunking, color coding, and access structure will be applied to human interaction with the screen display.

Also, guidelines will be outlined to suggest screen layout and grid systems, typography, color,

overall design guidelines and applications to course menus, query screens, touch buttons, and graphic-to-video relationships. In conclusion, two visual-based instructional constructs will be discussed.

The New Role of the Graphic Technologist

The GT's role includes the creation of an interface that provides a friendly medium between the learner and the electronic learning environment through the use of aesthetically pleasing, cognitively effective, and instructionally accurate screen displays. Specifically, the role of the GT can be defined as one who interprets, designs, and communicates information through the use of visuals such as graphics, video, text, icons, and other visual cues and devices.

The screen designer's previous responsibilities of creating mere visual enhancements is no longer acceptable as interactive systems and touch screens have been introduced. Designing for touch screen-user input demands that the GT compose the instructional interface of the entire IVD system. This interface must permit effective movement (mobility) through the course, such as within instructional modules, help sequences, visual databases, and resource directories, as well as from screen to screen. As result, the GT no longer creates isolated graphic stills and lifeless frames of text or graphics, but must begin thinking of the interface as a dynamic spatial and temporal relationship within an instructional framework.

The Interface: Screen Design, Course Mobility, and Integration

The instructional design of IVD courseware should reflect many of the same qualities of any other form of instruction or training, including the proper sequencing of content, as well as the necessary visual considerations. However, an important responsibility of the GT who is serving as a specialist in visual communications and visual factors is the relationship of three instructional dimensions within the IVD interface. The primary one, which is the focus of this paper, is screen design (Pictorial imagery). The second is extended system mobility, and the third is course integration.

Screen Design

A small group of researchers over the past decade have put forth significant efforts in the area of screen and human factors design (cf. Sherr, 1979; Shurtleff, 1980; and Galitz, 1981). They have proven that good screen design can represent a critical factor in the interface between man and machine. As a result, thoughtful utilization of text and graphics has proven to be: (1) significant in aiding insight and understanding the relation-

Tony Faiola, ALTA Associates (555 Metro Place North, Dublin, Ohio 43017), is a Graphic Technologist specializing in visual communications, visual factors, and authoring for computer-based instructional systems. Michael L. DeBloois Senior Consultant for ALTA Associates, is an instructional designer, educator, and lecturer.

ship between concepts, and (2) valuable in illustrating processes (Kearsley, 1982). Hence, the quality of screen design has shown to be a strong encouragement to improved performance when it maintains the interest of the user while also lowering the chances of confusion, eye strain, and fatigue caused by poorly designed information. From this we can gather that screen design should involve fundamental design principles and visual factors related to visual attributes and locations of textual and graphic elements. In the following section on screen design four main items will be discussed: grid system, typography, color, and general screen design guidelines.

Grid System

Given the characteristics of commercially available television receivers and monitors, a well designed screen display must rely on a grid system. The grid provides the GT screen designer with a tool for solving basic to more sophisticated visual problems and helps him or her be strategic. The grid also functions as a controlling principle and an ordering system. It directs how the GT controls the construction of the visual system and how sophisticated design problems will be solved.

Practically, the use of grids determines the variety of constant dimensions of space between text and graphic elements on the screen. The grid provides a consistent location for text, such as paragraphs, headings, tables, screen numbers, as well as graphic objects such as touch buttons, icons, and other forms of graphic representation. It should ultimately communicate a sense of planning and orderliness of design. The grid should provide a credible and a transparent framework to large and complex structures of instructional information. A well laid out grid provides a strong foundation and consistent guide to the GT in the creation of an effective interface. This in turn saves the viewer from erratic and confusing changes of course elements.

Establishing a usable grid requires foresight, thoughtfulness, and a sensitivity to all media involved, especially in respect to their relationship to the limitations of the viewing screen shape and size. Furthermore, human factors research findings show that a well-designed grid layout improves comprehension and shortens the reader's search time (Hartley, 1978, 1980). Though most of these experiments were done with print-based literature, the basic principles of typography and human performance remain constant in regard to legibility.

The GT can begin by drawing preliminary thumb-nail sketches on paper to map out the general screen location for all the various items involved—text, paragraphs, menu buttons. The GT should pay attention to negative space such as margins and areas around objects. The grid establishes an invisible framework on which to lay objects of text and graphics. Occasionally, the GT will confront unusual situations where conflicting portions of text or graphics will be necessary. In these situations, the GT must work with the exceptions, but if it is a constant conflict, he or she must consider adjusting the grid.

Both text and object location should be considered within a grid with the potential of background stills and moving video in mind. For example, established touch buttons that are too large, as well as located too far within the image area of the screen will interfere with video viewing. This is why you must know the kinds of elements which will eventually be brought together to form each screen display.

Questions must be asked such as: Which of the course frames will be mostly text or have a large number of graphics or video stills? Will certain portions of the screen remain unimportant to background information and be available for touch points, menu buttons, or information "boxes" and the like? The presentation of grid guidelines should be thoroughly considered before decisions are made about course mobility.

Typography

The need to effectively communicate large and complex systems of information by means of an IVD interface requires the GT to maintain specific typographical standards. Though IVD creates a new context for typography, the traditional principles of type should remain constant in application. While the establishment of a grid remains a fundamental prerequisite to sound screen design, typography plays a crucial role for verbal-graphic communication. With this practical view, the interface created by the GT can achieve a high degree of effective communication, while at the same time be attractive and friendly. Understanding the various aspects of typography means knowing some terms, as well as recognizing their relationship to visual factors.

Typeface/Font: The first term most thought of in relation to typography is typeface or font. However, typeface refers to the design of the letters of an alphabet, whereas font refers to a complete set of letters within an alphabet, upper and lower case and always of a single size. For example, Times Roman may refer to a typeface, but 10 point Times Roman may refer to a specific font.

In making typographical decisions for screen design, the GT is limited by the typefaces originally designed by the graphic or authoring software manufacturer, which are usually quite limiting. Software houses select typefaces based upon the technology at hand, trying to duplicate pre-established typefaces from the printing industry as close as possible. However, the commonly used 640 x 200 pixels screen (80 character line length) restricts their ability to produce really attractive typefaces as in computer typesetting. For example, Dr. Halo and Paintbrush + provides approximately 25 different fonts. However, because of the limitations of the 640 x 200 resolution designated by the graphic card, character forms still appear rather ragged in appearance.

Measurement/Spacing: Measuring type is an important aspect of good typography. Instead of using the traditional form of measurement called points and picas, the most accurate way to measure type on the screen is by pixels or bits, as some graphic software identifies them.

Some graphic software designate point sizes to their fonts, but quite often they are not accurate. This method also is not realistic because this kind of measurement cannot be used with the 640 x 200 pixel resolution as it can with traditional measuring methods in the printing industry. Pixel-based measurement can also be used for letter space, word space, and line space. Letter spacing refers to the space between letters, word spacing to the space between words, and line spacing to the space between the lines, sometimes called leading.

Without getting into too much detail, it is important to realize the impact that can be achieved through careful use of negative spaces in typography on the screen display. Furthermore, these applications of spacing should be regulated not merely because of the general need for good typography, but because of the characteristics of specific typefaces. In other words, certain computer generated fonts will require more modification, depending on their typeface and size.

Modifying negative spaces around type is another possible way to greatly improve legibility in type on the screen. Legibility here means the visual factors related to how well type can be seen and physically read with ease. This is especially true with headings and bodies of text, which already suffer because of poor screen resolution. However, this major hindrance can be partially overcome by the proper letter, line, and word spacing. It is suggested that for maximum ease of reading and legibility the amount of space between letters in a single word appear to the viewer as equal, though in reality they are not. The juxtaposition of certain letters creates an optical illusion of being closer or further away from each other than they actually are, so modifications are often necessary, i.e., the letters AQ QU. Notice that when graphic software brings text up on a screen it often does not account for this need for optical spacing; somewhat like a typewriter, which designates equal space to all characters, whether it is an I or W.

Arrangement: Another aid in achieving a higher standard of design and legibility in the IVD interface is the proper selection of text and heading typefaces and their arrangement on the screen. As you may have already experienced, only certain computer fonts and font sizes provide a clean and attractive image for text. As far as headings go, the proper modification of letter space is especially helpful to improve their legibility and can often increase attractiveness and clarity. Heading sizes can vary depending on their usage. Arrangement also refers to the justified, unjustified, or centered locations of text or headings.

Selection: Be selective in choosing typefaces. It is disconcerting to use more than two or three typefaces and sizes within one screen frame. With the existing limitation of legibly generated fonts, it is best to select at the time of the completion of the grid two fonts to be used throughout the IVD interface—designating specific sizes to each font, depending on their function. Often these decisions cannot be confirmed until the creation of a few screens has taken place; however the tentative establishment of these attributes is critical. It is also at this time that the selection of headings, text, and numerical colors are chosen. (This will be discussed in more detail in the section on color coding.)

Textual Cueing: The meaningful selection of fonts and sizes being related to a particular function of text is also related to textual cueing in instruction. The consistency in typographic attributes can establish and convey a very clear visual message to the viewer that he is now reading a certain subject or section of instructional content. For example, if for 20 screens, an eight pixel, yellow typeface has been used for questions only, whereas white of the same size was used for instructional text, a change from yellow to white would immediately confuse the learner. This principle of consistency should be strictly adhered to as to the responsibility it bears in assigning textual cues and messages to the user. This consistency in cueing should be applied to all usages of fonts, sizes, spacing, color, and any other kind of attribute. Quite often the only time one may find room to make a change is in the beginning of a new subsection of instruction to identify to the learner a major change in instructional content, system function or course branching. Both the subject of consistency and cueing related access structure will be discussed in more detail later.

Enhancements: All the preceding typography principles can also be applied to the use of rules, bullets, and underscoring. The over use or misuse of these typographical elements can greatly clutter up a screen, leaving an obtrusive matter of colors, spots, confusing lines, and distortion of text.

One final, rather attractive method of enhancing text is the use of an initial letter. This is to use a large letter at the beginning of a paragraph of a major section of course content. This form of embellishment has been used for thousands of years, notably by the scribes of ancient manuscripts such as the Bible, and if used tastefully can be a refreshing element that adds a touch of class to the introduction of any IVD system.

Legibility: Based on studies by CBT researchers (Caldwell, 1980; Faiola, 1984) in the area of typography and screen design, an emphasis should not be placed primarily on reading speed, but on the ability of the screen designer to make reading easier and at the same time maintain the interest of the reader in an IVD environment. Hence, "chunking" large portions of course materials into smaller units has been shown to improve visual clarity and result in improved retention of information in a complexly structure instructional framework.

The concept of "chunking" information is further supported by the research done by M.A. Tinker in the area of legibility of print. Since the 1920's, psychologist Tinker (1963) has proven that a text could be made more comprehensible and more functional through deliberate emphasis on form. In other words, through the careful layout and selection of textual attributes, the deliberate designing of text can improve the reader's comprehension. Based upon these rudimentary principles of text design, the GT can begin to consider some valuable practices for the design of screen displays.

Caldwell (1980) suggests that screen designers break large portions of course materials into discrete paragraphs and units. He also suggests double spacing text and the use of color coding to highlight particular words and sentences that require special attention. However, the regulated testing of such techniques is necessary to assure optimum legibility and functionality of text and graphics to the user.

Another study of visual factors has found that levels of legibility are higher when lower case text is used for screen design. The variation of shapes makes the screen more legible, primarily because words are perceived by shape and out-

line and not letter by letter. As a result, researchers have found that lower case text provides optimum levels of legibility to the reader for titles, subtitles and bodies of text.

Access Structure: The term "access structure" refers to the coordinated use of typographically signalled structural cues that help students to read texts using selective sampling strategies. Hatt (1976) suggests that it would be interesting to have a more comprehensive framework (i.e., an access structure) for evaluating and predicting the effect of text reading processes and behavior. Waller (1979) further adds that there is a link between behavioral research of reading processes and the formal analysis of a text structure.

A skeptic may be quick to ask "what does access structure have to do with typography in IVD?" An access structure is important for any high degree of understanding. Unless an IVD environment can maintain a systemized two dimensional typographical structure and logical sequence of visual events, the instructional concepts underlining that course will appear weak and fragmented—especially with the juxtaposition of graphics with audio/visual sequences. In fact, without adequate knowledge of cognitive processing and perceptual factors of textual information, the GT's typographical decisions in devising an access structure are based primarily on subjective feelings which are likely to lack consistency and precision. An access structure provides a variety of signals that are useful in helping learners discriminate among different contextual elements. When used properly and consistently, the reader is provided a signalled structure for easier access.

As a result, an access structure includes an overall textual framework of access signals composed of the inter-working of the grid system, textual cues, such as color, shape, size, etc., and the superstructure of an instructional design. With this in view, the utilization and reading processes of headings, lists, and a chunk of text is significantly more effective.

Visual Factors Research: Waller (1980) has stated that the study of typography as a functional component of written communication has a relatively short history. As a result, there exists a poor armory of concepts, definitions, and frameworks with which to discuss the issues. Because of this deficiency of formulated research and knowledge, typography has too often been considered by some as simply a decorative or tactile manipulation of graphic elements; not recognizing the didactic capabilities of this system of visual language for communicating within the social, industrial, and educational domains. To utilize typography as a functional and effective component of visual communication in IVD, one must draw some clear distinctions between mere cosmetic treatment and typographic research of visual factors.

The traditional techniques of typographical design and editing of CBT goes back to the creation of the textbook, which was often based on inaccurate and inappropriate assumptions on how books are read. With the addition of complex color graphics, icons and video images, the printed page and video screen are increasingly more difficult to compare. There is a primary need for quality research related to standards of empirical observation and by development of a well considered and tested conceptual framework for video screen-based typographical applications.

To date, the body of knowledge of screen design in this area is derived from experimental studies and is quite meager. Minimal attempts have been put forth to systematically apply the wealth of information and techniques available concerning the design of printed materials to screen design. Until now, systems designers mostly rely on inspiration—making screen design and system design decisions and problem-solving neither based upon visual factors testing, sound conceptual frameworks, or techniques of visual communication. As a result, most of today's screen displays desperately lack the conceptual and visual clarity as well as instructional sophistication necessary to create an effective IVD interface.

In a statement directly related to print-based typography, Waller (1976) stresses the importance of observing standards. His statements can also apply to the visual design of quality IVD and CBT in general. Until researchers can provide realistic screen design, performance standards to add to aesthetic standards, screen design will continue to be generally regarded as peripheral to the communication process.

Color

After viewing a green mono-colored monitor for many years, there is a tendency to be overwhelmed with opportunity when given a color enhanced monitor. There is always a phase of visual bombardment and then regression which transpires when first working with graphic software, especially if the designer has an untrained eye for color and design. Eventually the GT will observe that color selections and combinations that once charmed later on may send a different message. Unseasoned use of color selections and combinations may interfere with the real intention to communicate information. Misuse and misunderstanding of the use of color is common. This is not only in any number of software packages (from games to word processing), but also in computer-based learning environments, where accurate and efficient communication is the most crucial.

Color as a Powerful Tool: As with the effective use of any tool in IVD, color requires awareness of certain principles. Designers should also acquire a minimal knowledge of color theory and design, visual factors, and psychology. To effectively use color, the GT must understand precisely how it communicates. Color has the dynamic power to change the entire condition and atmosphere of an IVD interface and can motivate and evoke a large range of physiological responses. Color can aid memory and enhance the understanding of information. Today, many instructors underestimate the power of color—treating it as a toy for their entertainment while investing more than 90% of their time to other matters in IVD development. But for thousands of years artists and designers have used color to make impressionable statements, relying on it to convey feelings, thoughts and an entirely new realm of phenomena. The use of color in IVD can be just as involving or detrimental to the success of communicating course information.

Color Terminology: Being more complex than imaginable, color is a composition of three interacting aspects: hue, value, and chroma. The same innate principles of color can be applied to computer-based color displays which are a luminous (light-emitting) source, as with reflective color printed materials. Hue is what most of us usually refer to as color itself. The primary hues are blue, red, and yellow. These primary colors are those common denominators that, when combined together, produce all other color. Of course, by adding the element of value and chroma their quality can be altered. Value, also referred to as brightness, is simply the intensity of light, with degrees of intensity ranging from light to dark. For example, blue, red, and dark grey would be dark and yellow and white would be light. Chroma, also referred to as saturation, on the other hand, is directly related to the addition or subtraction of white to any color.

Color Interaction: After discussing terminology and theory of color, it is more crucial for the GT to gain an experienced knowledge of the interaction or relationship of color. The most direct way to understand these relationships is through a variable just as important as hue, value, and chroma; that is, contrast. Simply, contrast is based on relativity of one color with another. The better the contrast, the higher the degree of readability of text on a display. For example, using white or yellow text on a black or gray background improves readability; its opposite lowers the contrast, and thus reduces readability. Contrast can be created by selecting differences in hue, but also by varying value and chroma. In these cases when used properly, they tend to produce the effect of spatial differences, and shadowed or shaded areas, thus providing the GT with the illusion of depth.

Color Guidelines: Although we will not discuss color design, harmony, or theory in depth in this article, the reader should be aware of the need to attend to these tools and how they work together with the preceding ideas. The following statements are offered by technologists to provide some mapping points in the IVD interface and color applications.

a. Though the eye can perceive millions of colors, limit your color palette per screen to what the eye can actually keep track of at one glance. This usually is about six colors depending on the complexity of the screen design. In most cases, however, if the screen design only consists of text and one or two graphic elements like bars, buttons, or menus, the color selection should be limited to four colors.

b. Long-term users, with more screen display interaction time, are capable of perceiving and responding to a broader range of color and their coding relationships, so selections can be increased accordingly.

c. Consider carefully the selection of colors for all your visual devices such as touch screen, buttons, menus and titles, etc. Remember, these devices will be standing alone in a variety of backgrounds and visual environments.

d. Be absolutely consistent in all color choices. For example, if a dark blue background with white text is used for menus, keep them the same, unless your IVD environment is branched in such a way that a major subset of menus must be visually distinct from the rest.

e. In selecting color combinations make sure they are compatible. For example, complementary color pairs such as blue/orange, red/green, and violet/yellow should be avoided or used with extreme discernment, based on design specifications.

f. Using the range of grays is a safe way to provide a neutral background for the enhancement of two or three other colors. Gray is especially effective in the use of access devices such as touch buttons, which are necessary to the user, but should not be a distraction from other textual information or video.

g. Be more attentive to value than hue in your concern for legibility of textual information. For example, always select text on background colors which creates contrast; white on black, white on dark gray, white on dark blue, dark blue on light gray or white, black on light gray or dark gray.

h. Avoid background colors that are too high in value and chroma, especially if it is merely a title page with two to four words. The optical response to high value or chroma hinders the legibility of the text.

Color Coding Guidelines:

a. Use commonplace denotations in color selections for coding particular action-related cues. For example, red for danger, stop, caution, warning; yellow for yield, pause, consider, wait for a moment before going on; green for go, go to next screen, proceed, okay, all clear. Of course, culture boundaries, emotions, and social structures are also a major consideration in this category of color denotation.

b. Because of the poor color memory of some users, or even color blindness, major color coding items should be selected which make distinct hue, value and chroma differences. Perhaps in this case, color in combination with iconic images can provide a more accurate coding of access devices and cueing.

c. Consistency in color coding is also crucial. If one color means one thing maintain that system of denotation. Redundancy in coding can only improve visual communication of information. This is especially true in training the user to the variety of visual cues and access devices for smooth mobility through the instructional system.

d. When a quick response is necessary as a result of the course design, use colors with higher degrees of chroma.

The designer must be aware of the variables in color utilization and must experiment with numerous combinations of hue, value, and chroma before confirming his choice, reserving the option for later changes. Then, after preliminary color choices are made in cues and other devices, test the tentative selections on trial users for immediate feedback to ascertain legibility, readability, coding denotations, color functionality, balance, and overall usability and friendliness.

Color Coding: Research of this type provides a base for the existence of a larger spectrum of user association and response to color and color combinations. The GT must be intimately involved in the decision making process of color selection for color coding text and all other access or cueing devices.

He or she should avoid any overly simplified understanding and identification to the effects of a color statement in coding.

Research consistently suggests that the double spacing of text and the use of color coding to highlight particular words and sentences requires special attention. The regulated testing of such techniques is necessary to assure optimum legibility and functionality of text and graphics to the user. Examples of such research in screen legibility can be seen in the testing of visual search time on color-coded information displays and characters of four different dot-matrix sizes by Carter (1982) and Pastoor, Schwarz, and Beldie (1983).

From the preceding section on screen design the GT should note that the adequate knowledge from visual factors research remains crucial to optimal delivery of course content to the IVD user. \

General Screen Design Guidelines

With the aid of a compilation of materials on screen design, the more important CBT screen design concepts include: design prerequisites, design consistency, design standards, and design trade-offs (Galitz, 1981; Marcus, 1980).

Design Prerequisites

It can be dangerous to ask a graphic technologist what criteria he or she uses to determine the difference between good and bad instructional displays. Like politics and religion, this topic is fraught with potential conflict. The designer makes a variety of decisions based on the fundamental understanding of various conceptual strategies for articulating course materials and the following items:

a. Human factors principles (cognitive/behavioral psychology).
b. Basic understanding of graphic design principles.
c. Understanding of the instructional design which must be interpreted through an effective visual based interface.
d. Informal research conducted by the GT.
e. The GT's personal experience with IVD development.

Design Consistency

a. Touch Screen Procedures: Every time a specific screen location is used, the exact same predictable action must occur.
b. Screen layout:
   1. Consistent relationships, placement of identifying information such as titles, headings, instructional information, visual cues, and touch points.
   2. Consistent relationship between headings and captions.
   3. Consistent placement of common or frequently occurring information.
   4. Consistent physical characteristics, e.g., size, color, font/typeface, and other typographic attributes.

5. Predictable rhythm or action of timed elements in video and frame-to-frame.
6. Audio consistency in volume and pitch.
7. Appropriate selection of words to match predetermined readability levels.
8. Well balanced visual and audio media mix in motion, stills, audio segments, etc.

Design Standards: Standards Should Guarantee User:

a. Consistent appearance and procedural usage.
b. Visual clarity, legibility, and ease of usage.
c. Organizational integrity.
d. Cost effective media placement on video and floppy disk.

Design Trade-offs

a. Visual factors must always take precedence over machines, processing requirements and aesthetic considerations.
b. The completed design may necessitate a compromise while weighing the alternatives of accuracy, time, cost, and ease-of-use.
c. General: In properly analyzing a dilemma where a trade-off is warranted, the conceptual framework of course content and its relationship to human performance must take priority in all problem-solving of screen display layout of course materials.

Extended Course Mobility

When preparing an IVD learning environment that can effectively maintain the attention of the user, include the significant and intricate visual treatment of screen design as only half of a functioning part of a two-fold interface. That is, the environment consists of an integration of pictorial elements along with course structure and usability. In this section, two instructional models can be used to convey these concepts. The first is related to course mobility and the second to the course integration of content and structure. Later discussions will show how these two models are interrelated.

The first model is related to course mobility. The continual availability of a "Help" or "Next" option in each frame provides continual accessibility to extended course mobility via an expanded menu of options when "Help" is touched. The practical implementation of this tool provides the user with a high degree of movement within the learning environment. Though an instructional framework may be complex in nature, its complexity can be reduced by (1) a well ordered system of branching and (2) a complementary menu system that can provide easy and regular access to all key locations in the learning environ-

Figure 1

Figure 1

Helix Arrangement

Figure 2

Traditional Arrangement

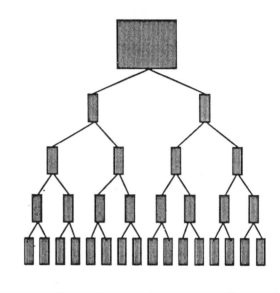

ment with an electronic bookmark placed for easy return. Figure 1 illustrates a downward blossoming helix which conveys the cyclical movement which is possible through a variety of access points. These points, illustrated by dots, represent frame locations. We choose a helix as opposed to the traditional pyramid-like branch, Figure 2, because it more accurately conveys the movement possible when key menus, such as "Help," are easily accessible to learners at the beginning of the course. Modules are presented which allow access of content at different level. The blossoming helix, as it descends at the base, conveys the broadening and scope of course content. At the same time, there is a constant running vertical rod through the center of the helix. This rod represents the continual accessibility of the "Help" menu.

The "Help" menu is the key system access point with three functions:

1. It constantly provides the student with a glossary, calculator, list of visual and graphic "Figures," or any other form of learning aid.
2. It constantly provides the student with immediate access to the main menu in which he is allowed instructional access related to course subjects, topics, modules or content.

3. It constantly provides a high degree of framework mobility by means of a "Menu Jumper." The "Menu Jumper" is a menu which consists of all the course menus and other key access locations throughout the environment. This could total 10-20 menus or locations, depending on the size and complexity of the course structure. Using the "Jumper," the student can quickly advance to or review any course segment and then return to the point of departure.

The difference between the main menu and the menu jumper varies. The former provides a superficial entry point into primary subsystem locations, macro chunks of information organized following an empirical design, where as the latter provides secondary and deeper accessibility into the course micro environment. This allows student movement not only to be cyclical, but also progressively broadening as he or she reaches the base of the descending helix.

Course Integration

In devising the visual model for the concept of extended course mobility, a further model evolved which more accurately conveyed a philosophy which the ALTA team attempts to employ while creating the interface for an instructional framework.

With extended course mobility in full operation, the construct of course integration spontaneously

Figure 3

Course Integration Model

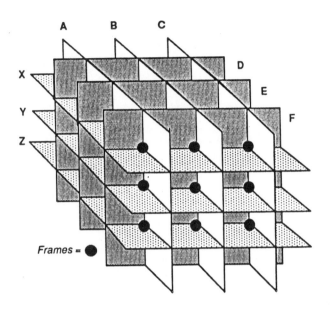

Frames = ●

falls into play providing a learning environment which is composed of an interwoven network of pictorial and mobile transactions. In Figure 3, we see three dimensions and numerous intersecting apexes:

1. Pictorial devices = text, graphics, icons, related attributes
2. Extended course mobility = menus, help, next, touch buttons
3. Subject matter terms = facts, concepts, rules, solutions

Pictorial devices comprise frames (A,B,C) of visual stills of graphics, text and their attributes. Extended course mobility (D,E,F) includes the provision of dynamic movement. As mentioned in the previous section, this movement is accessible through access devices, e.g., "Help," "Next" and video control buttons. Content elements include the subject matter being taught (X,Y,Z).

Together, these components integrated into a well designed access structure make an effective, efficient, practical and friendly IVD interface. Each frame (represented by a dot) must be created as an integration of the three intersecting planes (A,B,C; D,E,F; and X,Y,Z).

Conclusion

In this article, we have attempted to provide those involved in interactive video courseware development our view of the role of the graphic technologist. In addition, some necessary criteria are presented for the GT's role as a more critical and effective part of the IVD team. The responsibility of this role will vary from company to company and from project to project. However, in principle, the focus on visual factors and design as well as course strategies is necessary. We hope these aids and guidelines will assist other courseware developers to see and apply more rigorously the rules of good visual communication, keeping in mind the cognitive factors that affect student interaction and comprehension. □

References

Braden, R. Visuals for Interactive Video: Images for a New Technology (With Some Guidelines). *Educational Technology*, May 1986, 18-23.

Caldwell, R.M. Guidelines for Developing Basic Skills Instructional Materials for Use with Microcomputer Technology. *Educational Technology*, October 1980, *21*(10), 7-12.

Carter, R.C. Search Time with a Color Display: Analysis of Distribution Functions. *Human Factors*, 1982, *24*(2), 203-212.

Crouwel, W.T. Typography: A Technique for Making a Text Legible. *Processing of Visible Language 1* by Kolers, P.A., Worlstand, M.E., and Bouma, H. (Ed.) New York: Plenum Press, 1979, 151-164.

Faiola, T. Evaluative Study of Text from a Graphic Designer's Perspective. The Ohio State University, Department of Industrial Design, Columbus, Ohio, 1984.

Faiola, T. Design Guidelines for Text Preparation: Writing Text Macros for Information Systems. The Ohio State University, Department of Industrial Design, Columbus, Ohio, 1984.

Faiola, T. A Syntactical Approach of Inquiry into Complex Text Design. The Ohio State University, Department of Industrial Design, Columbus, Ohio, 1984.

Faiola, T. A CAI Overview and Prototype Document. The Ohio State University, Department of Industrial Design, Columbus, Ohio, 1984.

Galitz, W.O. *Handbook of Screen Format Design.* Wellesley, MA: Q.E.D. Information Sciences, Inc., 1981.

Hartley, J. *Designing Instructional Text.* London: Kogan Page, 1978.

Hatt, F. *Reading Process: A Framework for Analysis and Description.* London: Bingley, 1976.

Kearsley, G.P., and Hillelsohn, M.J. Human Factors Considerations for Computer-Based Training. *Journal of Computer-Based Instruction*, May 1982, *1*(4), 74-84.

Krilof, H.Z. Human Factor Considerations for Interactive Display Systems. *User-Oriented Design of Interactive Graphics Systems.* ACM/Workshop, 1976.

Marcus, A. Computer-Assisted Chart Making from the Graphic Designer's Perspective. *Lawrence Berkeley Laboratory*, Tech. Report LBL-10239, April 1980, and *Computer Graphics*, July 1980, *14*(3), 247-253.

Martin, B.L. Aesthetics and Media: Implications for the Design of Instruction. *Educational Technology*, June 1986, *26*(6), 15-21.

Pastoor, S., Schwarz, E., and Beldie, I.P. The Relative Suitability of Four Dot-Matrix Sizes for Text Presentation in Color Television Screens. *Human Factors*, 1983, *25*(3), 265-272.

Rayner, K. Visual Cues in Word Recognition and Reading: Introduction. *Visual Language*, February 1981, 125-128.

Sherr, S. *Electronic Displays.* New York: John Wiley, 1979.

Shurtleff, D.A. *How to Make Displays Legible.* La Mirada, CA: Human Interface Design, 1980.

Tinker, M.A. *Legibility of Print.* Ames: Iowa State University Press, 1963.

Waller, W. Typography for Graphic Communication. *Institute of Educational Technology*, Open University, Milton Keynes, England, May 1976, 3-16.

Waller, W. Typographic Access Structure for Educational Texts. *Processing of Visible Language*, Vol. 1. New York: Plenum, 1979.

Design Factors for Successful Videodisc-Based Instruction

Greg P. Kearsley and Jana Frost

This article discusses the factors associated with the design of effective videodiscs for instruction. First, though, examples of outstanding videodiscs are described and the evidence regarding the effectiveness of videodisc is summarized. Some of the important design factors discussed include: interactivity, visual design, organization, active participation, metaphors and models, personality, and a team approach.

Some Outstanding Videodiscs

In a recent book (Daynes and Butler, 1984) over 200 different videodiscs are listed under the instructional category. Most of these discs have been made in the past three or four years. While this production level may pale beside the number of texts or videocassettes or computer programs made for instructional purposes in the same period, it nonetheless illustrates that a great deal of activity is going on in the instructional videodisc field.

Perhaps the most intriguing instructional videodisc to date is the Cardio-Pulmonary Resuscitation (CPR) learning system developed by David Hon for the American Heart Association and marketed by Actronics. This system teaches the psychomotor skills needed to administer CPR using an instrumented mannequin and a two-monitor system with light pen input. The system gives the learner detailed feedback in terms of correct hand placement, pressure, rhythm, etc. CPR skills are difficult to learn; this system shows that videodisc can tackle difficult teaching problems.

Another "hands-on" type of videodisc system is the MK-60 Tank Gunnery Trainer developed by Perceptronics for the U.S. Army. This system features a mockup of the actual firing station in an M-60 tank, and the videodisc is used to provide realistic sequences for targeting and shooting. Similar systems employing videodisc have been developed for a number of weapons systems covering both operation and maintenance of the equipment.

The Tacoma Narrows videodisc was one of the first discs to be marketed commercially by a publisher (John Wiley & Sons). This videodisc was developed by R.G. Fuller (1984) and colleagues at the University of Nebraska and teaches basic principles of harmonic motion in physics instruction using the collapse of the Tacoma Narrows bridge as a case study. The disc features three levels of instruction: qualitative explanations using graphs, numerical calculations, and algebra.

The National Kidisc produced by Optical Programming Associates in 1981 was one of the first widely available educational videodiscs. It was aimed at home use rather than schools and contained 28 different activities for young children. Optical Programming Associates has released a number of other similar "edutainment" discs including the "History Disquiz" and "Fun and Games."

The Action Code system originally developed by Perceptronics and now marketed by the ICS/Intext Division of National Education Corporation provides approximately 400 hours of electronics training covering basic skills through advanced troubleshooting. The Action Code system consists of a microcomputer-controlled videodisc player, optical scanning wand (bar code reader), and bar-coded text books. When the student passes the wand over the bar codes in the text, still frames or video sequences are displayed as explanations, examples, problems, or feedback.

The Diagnostic Challenges videodisc produced by WICAT, Inc., for the pharmaceutical company Smith, Kline, and French, provides a diagnostic simulation for gastroentrologists. Physicians are presented with patients and options to look at medical histories, lab tests, and X-rays, and may even conduct an endoscopic examination. Miles Laboratories subsequently has developed a similar series of diagnostic medical videodiscs.

IBM has used videodisc extensively for training customers to use their computer products. For example, a network of approximately 30 Guided Learning Centers equipped with videodisc players has been established to teach users how to operate various office and data processing systems (E&ITV, 1981). IBM has also produced videodiscs for large software systems such as its COPICS integrated manufacturing system. Wang has devel-

Greg P. Kearsley is Chief Scientist, Courseware, Inc., 10075 Carroll Canyon Road, San Diego, California. Jana Frost is a project manager and instructional designer at Maritz Communications, St. Louis, Missouri.

oped a customer training videodisc for one of its automated office systems, and several companies have developed customer training discs for personal computer software programs.

One of the more interesting applications of videodisc is surrogate travel pioneered by Nicholas Negroponte and colleagues at MIT. These techniques allow a user to take a "trip" through some domain (geography, space, a piece of equipment, inside the body) controlling progress via a joystick. Surrogate travel has been used to familiarize intelligence officers and soldiers with foreign locations prior to their arrival. HumRRO has applied these techniques to Army training in areas such as navigation and stress reduction (HumRRO, 1982).

A final example is the JOIN (Joint Occupational Information Network) System developed by the U.S. Army. JOIN is an automated information-management system for recruiting. A microcomputer-controlled videodisc is used to show potential recruits the nature of various military specialties as well as to train recruiters. The microcomputer also generates the forms and reports associated with recruiting activities. Similar systems could be used for civilian counseling activities or other information-management applications.

The examples just described serve to illustrate the range and variety of instructional videodisc applications. Other current projects include art instruction, the teaching of sign language to the hearing impaired, management skills training, industrial machinery and fabrication training, robotics troubleshooting instruction, courtroom practices, and many others. Kearsley (1983) and Schneider and Bennion (1981) provide descriptions of many projects completed to date. Next comes the question of effectiveness.

Effectiveness of Videodisc

A large amount of evidence has now been accumulated regarding the effectiveness of videodisc-based instruction. This evidence strongly supports the view that videodisc is a highly effective technology for teaching and learning. While few of these studies are methodologically rigorous, they do provide useful indicative information for assessing the worth of instructional videodisc.

The largest collection of evaluation studies has been conducted by the U.S. Army. An experiment at the U.S. Army Air Defense School compared classroom, computer, and interactive videodisc simulations for maintenance training on the HAWK missile control system. The results showed that the interactive videodisc group took half the time to solve repair problems (Kimberlin, 1982). In a study conducted at the U.S. Army Signal Center, practice with actual equipment was compared to videodisc

simulations for maintenance of a satellite receiver. The videodisc group mastered the skills in 25 percent less time (Ketner, 1982). Similar results were obtained with a videodisc for teaching paramedical and basic nursing skills developed by the U.S. Army Academy of Health Sciences: the training took one third less time with the same levels of achievement (Manning et al., 1983).

Another Army study compared videodisc instruction with programmed text and role playing for training officer leadership and counseling skills. Videodiscs were found to be superior to the other two modes on tests of understanding (Schroeder, 1982). One interesting evaluation of videodisc in the context of Army training was conducted by Holmgren et al. (1980). In this study, existing training extension course materials in film/slide format were compared with two videodisc versions: an exact duplicate and one in which review frames had been added. The results showed that all groups did about the same, indicating that videodisc has no inherent advantages given content originally developed for other media.

A large number of evaluation studies have been conducted in education contexts. Some of these studies are summative in nature while others are formative. For example, Bunderson et al. (1981) compared student achievement in biology with videodisc versus classroom instruction at the college level. The videodisc group scored significantly higher on the posttest and required about 30 percent less study time. Boen (1983) also compared classroom instruction versus videodisc for teaching test-taking skills with university students and found videodisc instruction superior. Glenn et al. (1984) conducted a pre/posttest study with an economics videodisc and showed significant learning gains.

In a field test conducted with a health education videodisc for elementary children, 94 percent of the pupils indicated that they learned a lot by using the videodisc (Kirchner et al., 1983). This study suggested that both teachers and youngsters were able to learn to use videodisc with minimal training (although the more training, the better it was used). Similar outcomes were obtained in a study by Daynes et al. (1981) with a gymnastics videodisc for elementary physical education teachers. This study identified a number of strategies developed by the teachers for use of the videodisc with an entire class.

Hon (1983) reports on the effectiveness of the CPR videodisc mentioned earlier. Fifty students took the CPR course from an accredited live instructor and fifty took the course from the CPR videodisc system. Both groups of students were evaluated by another instructor. Three times as many of the students who had the videodisc training passed as those who had taken the instructor-based training. In addition, *all* videodisc students took less time than *any* of the students in the regular class.

A number of studies of videodisc effectiveness have been conducted in the domain of corporate training. A study by American Bell on the effectiveness of a product training videodisc in branch office training indicated that the videodisc was more effective than classroom, text, slide, or video*tape* training (Goldberg, 1983). The use of a pilot videodisc for railroad safety training was received very positively by employees (Fedewa, 1983). Duelfer (1983) describes the comparison of a videodisc with instructor-led courses for word-processing training. The videodisc students performed better on a posttest and reported participating more in the training than with classroom instruction.

Hershberger (1984) conducted a study which examined the effects of interactive video on corporations. She summarized her results in terms of organizational impact, employee impact, and trainer impact. She found that interactive video training stimulated employee involvement in training; encouraged more employee independence and self-sufficiency; and increased the esteem and professional status of trainers.

To summarize, the available evidence suggests that videodisc is a highly effective instructional medium across all types of educational and training applications. Typically, students who learn via interactive video achieve better test scores with less training time required. Videodisc is well accepted by students, instructors, employees, and managers. In the hands of talented and experienced instructional developers, videodisc has been demonstrated to be one of the most powerful instructional technologies currently available. We now turn to consider the design factors that make videodiscs effective.

Design Factors for Effective Videodisc-Based Instruction

Interactivity

There is one fundamental concept that ties together all of the design factors underlying videodisc: interactivity. We address this general concept before discussing more specific design factors.

Videodisc is an inherently interactive medium. However, the nature of the interaction possible depends upon the capabilities of the videodisc delivery system. A well-accepted categorization has been developed that distinguishes three levels of videodisc systems. Level 1 involves a videodisc with no programming of any form. An inexpensive "consumer" type player can be used for Level 1 videodiscs. Level 2 involves a videodisc which contains a program on the disc and requires a player with a built-in microprocessor to read the program. Level 3 consists of a videodisc player controlled by a program stored on a separate computer.

The interesting point is that all three levels are interactive but in different ways. Level 1 interactivity allows the user to manually branch to locations on the videodisc as well as speed up or slow down the video presentation rate and switch audio channels. It is also possible to create different feedback sequences using the step and auto-stop (chapter) features of players.

With a Level 2 system, it is possible to have true branching for answer sequences or menu selections. However, because the programming languages provided with these players and the amount of memory is fairly limited, the response analysis possible is minimal. In contrast, a Level 3 system, which relies upon an external microcomputer for processing and storage, is not limited in this fashion. Furthermore, Level 3 systems provide the capability to store responses and generate graphics.

While it is clear that more sophisticated forms of interactivity are possible with Level 3 systems relative to Level 1 and 2 systems, it should not be assumed that Level 1 and 2 systems are inferior for all instructional applications. The large-scale implementations of videodiscs for automobile dealer training by GM and Ford utilize Level 2 systems. Steinbicker (1984) describes a highly effective use of a Level 1 system for manufacturing training at Honeywell. One of the virtues of Level 1 systems is their relative simplicity of design and use. Simplicity and usability are important design factors.

On the other hand, to achieve the sophisticated interactive capabilities exhibited by some of the outstanding videodisc projects described earlier, a Level 3 system is required. This complicates the equipment requirements and significantly raises the costs of videodisc development. Given the additional equipment and resource considerations associated with Level 3 systems, good justification should exist for using Level 3 over Level 1 or 2.

From a design point of view, all forms of interactivity are based upon branching. Branching occurs through the use of two basic conventions: a multiple-choice or free response. In either case, branching to another part of the disc occurs when an option is selected or typed in.

Traditional multiple-choice conditions use menus or multiple-choice questions. Too often we develop these in unimaginative ways. Videodisc technology allows us to expand upon these conditions. For example, the use of a touch panel makes it possible to select sequences of choices or to have "stop-action" options. The use of simulation and peripheral devices such as a joystick or a mannequin sets up unique option conditions for the designer.

In developing a design it is easy to get hung up with unnecessary branching sequences that never get used by the learner. There is also the tendency to underestimate learner needs with too few branches or grow tired of creating specific feedback messages and use standard lines over and over. Obviously, disc space is saved but quality is often sacrificed. Feedback messages can be presented in a variety of ways—audio, motion sequence, animated characters, beeps, etc.

Reduced to their basic nature, design considerations for interactivity have to do with the representation, specification, and programming of branching instructions and feedback messages. From this it follows that a good understanding of branching structures is a critical design characteristic for videodisc.

Visual Design

Videodisc is an inherently visual medium. The impact and effectiveness of a videodisc will depend heavily on the quality of the visual design. The same principles that apply to good design in film and television production also apply to videodisc. Hoekema (1983) argues that a higher quality of video production is needed for videodisc than other instructional media. This argument is based on the fact that videodisc can be examined on a frame by frame basis and that people *expect* "broadcast quality" on television images.

Animation is an important visual capability of videodiscs. This can be created via traditional film techniques or computer generated graphics. Furthermore, with Level 3 systems computer-generated graphics can be either a video component or produced by the computer in real time. The capability to generate the graphics means that they can be based upon user input. When both the video "background" and computer-generated "foreground" use the same animation format, the visual effect is very powerful. Many arcade games exploit this capability.

An important use of visual design is to direct the students' attention and stimulate interest. Fleming and Levie (1978) emphasize that human perception is very selective. A designer who can predict what the audience will selectively perceive in a message will be the most successful.

Videodiscs usually involve a large number of still frames. It is relatively easy to list a set of factors that will lead to good visual design for still frames. These include:

- Present one idea per screen.
- Avoid cluttering the screen with too much information.
- Use a small number of type styles and sizes to highlight ideas.
- Use color *sparingly* for cueing or emphasis and avoid extreme contrasts.
- Organize information functionally on the screen as much as possible to reduce confusion and unnecessary cognitive processing.
- Use graphics and visuals instead of text whenever possible.

This small number of principles, if followed, will consistently result in higher quality videodiscs.

To summarize, the quality of video sequences, animation, and still frames will determine the visual impact of a videodisc and hence contribute a great deal to instructional effectiveness. Knowing how to blend these major visual components together calls for a high degree of technical expertise and artistic talent.

Organization

One of the implications of the storage capability of videodisc is that it is quite easy to get lost and not be able to find specific instructional sequences. A great deal of attention must be paid to the organization of information on the disc and to providing "navigational" aids.

Information on videodiscs is usually organized in some hierarchical format and accessed by one or more levels of menus. Some discs feature an index like a book index, which lists the frame number(s) for particular topics. It is highly desirable to give each still frame a title or label and indicate its sequence number.

Space on a disc side is major constraint to any instructional design for videodisc. The best laid design plans may not fit the space available. Part of the problem is remedied by sound-over-still capability and graphic-text-overlay capability. One consideration is to have the program on several disc sides. The continuity of the program is maintained by replicating program segments on each disc so that the student doesn't have to reload a different disc for certain information.

Another important organizational consideration is a user interface, which is suitable for both first-time and experienced users. All discs should have a "default" path, which allows new users to get to the "meat" of the instruction quickly. Menu selections and fixed sequences are usually used to achieve this "streamlining." Many finer points of the videodisc will not be experienced by the user initially. However, as learners become more comfortable using the videodisc, they will want to take shortcuts and explore more complex learning strategies.

Many videodiscs provide learner control options, which allow the student to back up, step ahead, get help, return to a main menu, etc. These options provide the experienced user with ways to take shortcuts or get more detail. They are also important in terms of allowing the learner to feel in control of the system (rather than vice-versa). In addition, learner control options are a critical aspect of achieving a high level of student participation in the videodisc.

Instruction should cover operation of the videodisc as well as how the material is structured. First-time users may need to be presented with a game exercise that motivates them to operate the system while losing their fear of the technology. More importantly, many users of interactive videodisc learning systems may have little experience with self-study learning and need "learning to learn" instruction.

Active Participation

Videodisc should produce a high level of student participation and involvement. There are a number of factors which affect this, including the kind of user control provided, the relevance of the instruction, and the instructional strategy used.

Videodisc should allow students to control their pace, the instructional strategy (e.g., tutorial, simulation, test, etc.), and the "depth" of the instruction. The more control students feel they have, the more involved they will be in the instruction. A good way to engage the learner in the instruction is to make the learner a character in the scenario. This is one of the reasons why game and simulation strategies are usually very effective. Asking the student to take the role of teacher or consultant also achieves this effect.

Fleming and Levie (1978) point out that people are attracted to novel information presentations. With so many media presentation alternatives available, the videodisc designer can develop very novel and stimulating presentation sequences that capture the student's attention and involvement. Novelty is one of the distinctive features of games and simulations that results in a high level of active participation.

The relevance of the instruction is another factor which affects the degree of active participation. "Know your audience" is a critical piece of advice for designing an effective videodisc. Videodisc is capable of accommodating different program designs for different audience types. Besides age, educational level, race, and language, there are other aspects to consider, such as what type of medium or learning style is preferred by the student. Videodisc technology can provide straight video, computer-based instruction, or an integration. We can also present linear programs or a highly interactive simulation.

It is important to collect data in order to do a good job at customization. For instance, designing a videodisc for training requires a close look at the work environment to see what can be simulated. This may require a thorough job/task analysis and careful observation of the typical training approach (e.g., apprenticeship, trial and error, modeling, etc.).

There are many benefits of developing a videodisc which simulates the job environment, including opportunity for realistic practice, quicker knowledge or skill acquisition, and direct transfer of learning to job performance. Bunderson et al. (1981) describe a design methodology for capturing job performance with interactive videodisc.

Simulation is the preferred instructional strategy for the design of videodiscs. Simulation can teach problem-solving, decision-making, and judgmental skills, which are not easily taught via other means. Videodisc can encompass all types of simulation, including equipment simulations, diagnostic simulations, and simulations of interpersonal interaction. Furthermore, videodisc can be used for performance-oriented testing where the person demonstrates the skills rather than just answering questions about them.

Videodisc can also play an important role as a multimedia learning aid that a student uses to solve problems. For example, a disc to teach equipment maintenance can contain troubleshooting suggestions, schematics, testing procedures, assembly/disassembly demonstrations, etc. Similarly, a sales disc could contain customer scenarios with sales strategies, product information, information about selling to specific markets, techniques for handling objections or closing the deal, etc. In this role, the videodisc becomes an electronic encyclopedia.

Metaphors and Models

Many videodisc designs are based upon an implicit or explicit metaphor or model that provides an overall coherency and integration for the disc. Most Level 1 videodiscs are based on the model of a book with pages (i.e., frames) and chapters. Hoekema (1983) discusses the metaphor of "obedient servant" to describe the appropriate tone of response messages. Some videodiscs, such as the Tacoma Narrows disc, use a "detective" strategy in which the learner must solve a problem using the information available on the videodisc.

Clanton (1983) discusses general metaphors that

can be used to teach the use of computer systems. Functional and operational metaphors explain a new concept based upon the workings of something familiar. Organizational metaphors focus on how people use location and distribution of information to categorize new concepts (e.g., the "filing cabinet" model). Integrating metaphors allow separate functions to share a common conceptual model (e.g., the "desk top" model). These metaphors and others can be used for videodisc design strategies.

Some models have been specially developed for videodisc. For example, the surrogate travel and interactive movie methods pioneered at MIT are novel to videodisc. On the other hand, the model underlying the CPR videodisc system is the familiar one of coaching—although its implementation is unique. It seems likely that designers will invent new metaphors and models for interactive videodisc as well as adapting other, existing ones.

Personality

The design factors discussed so far have been relatively objective and straightforward. However, there is another factor which is critical but difficult to define. It is often referred to as the "personality" of a videodisc. Personality refers to the mood or affective state created by the videodisc. It is a collective function of the soundtrack, narrative, visuals/graphics, type of learner involvement, and instructional strategies employed. Videodiscs can create a learning environment which is prescriptive, supportive, authoritarian, relaxed, etc. Videodiscs can strongly influence the attitudes of learners depending upon the design of this factor.

We normally use instructional objectives to specify the intended outcomes of instruction. We seldom state affective outcomes. Yet, in many cases, the affective outcomes (i.e., attitude change) are important aspects of the training. Videodisc provides a powerful medium for achieving affective outcomes. This is closely related to the personality projected by the videodisc. Hence, it becomes an important dimension of design.

Team Approach

The last design factor we want to mention is the significance of a team approach with videodisc. It is generally accepted that a single individual cannot be the sole designer for a videodisc. The design of an instructional videodisc requires a number of different kinds of expertise not likely to be found in one person: instructional design, video production, graphics, and computer programming. Each member of a videodisc design team should be an expert in his or her field as well as having a general working knowledge of videodisc technology. This team of experts brings a knowledge of the techniques and ideas that have worked effectively with other media and extends them to fit the capabilities of videodisc.

The need for a team approach to designing videodisc has a number of implications. First, a team approach makes the design stage more expensive and time-consuming than for other media. Secondly, the management of the design activity is more complex because good coordination among team members is needed. Thirdly, the personalities of the team members must "click" if the videodisc is not to have a fragmented character.

A design team does not naturally fall into the dimensional thinking that is required for good videodisc design. Believing a videotape writer with no computer experience can write for a Level 3 videodisc is somewhat unrealistic. Everyone needs some type of orientation or training in order to design good interactive instruction. This orientation should start with basic concepts of videodisc design. If clients are involved, it is imperative that they receive a similar orientation/training.

Conclusions

We have discussed some of the major design elements of effective videodisc. All of these factors are important and collectively contribute to the development of good-quality videodiscs. DeBloois *et al.* (1982) discuss additional design principles for videodisc.

There are dozens of other design considerations to be taken into account in designing a videodisc. These include:

- the appropriate ratio of motion, animation, and still frames;
- when to use narration versus text;
- when to use existing versus newly created materials;
- studio versus location shooting of video sequences;
- the best color combinations to use in still frames;
- what input modes to use (e.g., keyboard, joystick, touch, keypads, bar codes, etc.);
- the use of one or two screens (separate or overlaid displays);
- how to use videodisc with other media (e.g., printed texts) and when to incorporate other media into videodisc; and
- the importance of interchangeability across different disc players/computers.

Almost all of these considerations will be made on a case-by-case basis for a given project and application.

Throughout this article we have ignored the real-world constraints of limited time and money, which strongly influence what and how much can be accomplished. It is critical that videodisc designers have a good grasp of what level of quality can be achieved given certain budgets and time frames. Once the budget and schedule are set, they become part of the design challenge to achieve the best results with the time and money available.

It is important to keep in mind that videodisc is only one form of instructional delivery. We must be careful not to lose sight of our goal—to improve teaching and learning. Simply adopting videodisc technology without also embracing the appropriate instructional design methodology is not likely to result in success. If designers pay attention to the factors discussed in this article, their videodiscs are likely to be more effective. □

References

Boen, L.L., Teaching with an Interactive Videodisc System. *Educational Technology*, 1983, *23*(3), 42-43.

Bunderson, C.V., Olsen, J.B., and Baillio, B. *Proof-of-Concept Demonstration and Comparative Evaluation of a Prototype Intelligent Videodisc System.* NSF Final Report. WICAT Learning Design Laboratories, Orem, UT, January 1981.

Bunderson, C.V., Gibbons, A.G., Olsen, J., and Kearsley, G. Work Models: Beyond Instructional Objectives. *Instructional Science*, 1981, *10*, 205-215.

Clanton, C. The Future of Metaphor in Man-Machine Systems. *Byte*, December 1983, 263-279.

Daynes, R., Brown, R.D., and Newman, D.L. Field Test Evaluation of Teaching with Videodiscs. *E&ITV*, March 1981, 54-57.

Daynes, R., and Butler, B. *The Videodisc Book.* New York: John Wiley and Sons, 1984.

DeBloois, M.L. (Ed.) *Videodisc/Microcomputer Courseware Design.* Englewood Cliffs, NJ: Educational Technology Publications, 1982.

Duelfer, A., Interactive Videodisc in Office Automation Training. *Proceedings of the Fifth Annual Conference on Interactive Videodisc in Education and Training.* Society for Applied Learning Technology. Warrenton, VA, 1983, 48-51.

E&ITV. How IBM Uses Videodisc for Customer Training, March 1981, 31-33.

Fedewa, L.J. Safety Training for Railroad Operating Employees. *Proceedings of the Fifth Annual Conference on Interactive Videodisc in Education and Training.* Society for Applied Learning Technology. Warrenton, VA, 1983, 26-29.

Fleming, M., and Levie, W.H. *Instructional Message Design.* Englewood Cliffs, NJ: Educational Technology Publications, 1978.

Fuller, R.G., The Tacoma Narrows Videodisc: A Personal History. In R. Daynes and B. Butler (Eds.), *The Videodisc Book.* New York: John Wiley and Sons, 1984.

Glenn, A.D., Kozen, N.A., and Pollack, R.A. Teaching Economics: Research Findings from a Microcomputer/Videodisc Project. *Educational Technology*, 1984, *24*(3), 30-32.

Goldberg, M., Survey Shows Disc Training Works Well for American Bell. *E&ITV*, June 1983, p. 57.

Hershberger, L., How Interactive Training Affects Corporations. *E&ITV*, January 1984, 64-65.

Hoekema, J. Interactive Videodisc: A New Architecture. *Performance and Instruction*, 1983, *22*(9), p. 6.

Holmgren, J.E., Dyer, F.N., Hilligoss, R.E., and Heller, R.N. The Effectiveness of Army Training Extension Course Lessons on Videodisc. *Journal of Educational Technology Systems*, 1980, *8*(3), 263-274.

Hon, D., The Promise of Interactive Video. *Performance and Instruction Journal*, November 1983, 21-23.

HumRRO. Instructional Applications of Spatial Data Management. *Videodisc/Videotex*, 1982, *2*(2), 191-197.

Kearsley, G. Instructional Videodisc. *Journal of the American Society for Information Science*, 1983, *34*(6), 417-423.

Ketner, W.D. Videodisc Interactive Two-Dimensional Equipment Training. *Proceedings of Conference on Videodisc for Military Training and Simulation.* Society for Applied Learning Technology, Warrenton, VA, 1982, 18-20.

Kimberlin, D.A. The U.S. Army Air Defense School Distributed Instructional System Project Evaluation. *Proceedings of Conference on Videodisc for Military Training and Simulation.* Society for Applied Learning Technology. Warrenton, VA, 1982, 21-23.

Kirchner, G., Martyn, D., and Johnson, C. Simon Fraser University Project: Part Two: Field Testing an Experimental Videodisc with Elementary School Children. *Videodisc/Videotex*, Winter 1983, *3*(1), 45-58.

Manning, D.T., Balson, P., Ebner, D.G. and Brooks, F.R. Student Acceptance of Videodisc Programs for Paramedical Training. *T.H.E. Journal*, November 1983, 105-108.

Steinbicker, W.J. Walking with the Disc. *Video Manager*, June 1984, 24-25.

Schneider, E.W., and Bennion, J.L., *Videodiscs.* Englewood Cliffs, NJ: Educational Technology Publications, 1981.

Schroeder, J.E., U.S. Army VISTA Evaluation Results. *Proceedings of Conference on Videodisc for Military Training and Simulation.* Society for Applied Learning Technology. Warrenton, VA, 1982, 1-3.

Educational Strategies for Interactive Videodisc Design

David Deshler and Geraldine Gay

Purpose

The purpose of this article is to present a number of educational strategies which were derived through scanning, synthesizing, and simplifying implications from a wide variety of learning theories, and to make practical suggestions for interactive videodisc design. The list is not exhaustive, but merely suggestive of the many alternative ways to hypothesize and to structure interactive video activities based on assumptions about learning in behavioristic, cognitive, and humanistic learning traditions.

In this article, we make the following assumptions:

1. Videodisc systems are most likely to be effective if learning theories inform their design.
2. Application of videodisc systems should not be limited to lower levels of behavioral or cognitive functioning. The visual and symbolic capabilities of videodisc make it possible to enhance more complex levels of cognitive learning as well as to aid learners in criticism of psychological, political, or social phenomena, and in reflection on their personal attitudes, values, and aspirations.
3. The capabilities of interaction (control, feedback, pace) provide additional features to the traditional use of visuals in education. The list of educational strategies provided in this article is not new to educators who have emphasized learning through visualization. However, videodisc/microcomputer technologies can expand access and multiply the flexible use of visual and audio material.
4. Combinations of videodisc and microcomputer technologies may give rise to addi-

David Deshler is Associate Professor of Education, Cornell University, Ithaca, New York. Geraldine Gay is Adjunct Assistant Professor of Education and a research and development specialist, Cornell University Computer Services.

tional instrumentation of learning theory research, since a broad range of behavioral, cognitive, and affective operations can be advantageously designed and tested together.

Promise of Videodisc Systems

It has long been recognized that visualization is an important part of learning. What is unique about interactive video systems is the capability to link various modes of presentation for learners and to have many ways of accessing knowledge. Coupled with the capacity of computers to represent and manipulate graphics, the combination of realistic and graphic images represents an unprecedented opportunity for the learner to bring these together. Students are able to access video images, graphics, and text, and to use these representations of material to clarify understanding (Salomon, 1979).

Most computer-based learning programs are predominantly oriented to words and text. Graphic and visual representations greatly enhance instruction about spatial relationships, and about objects or procedures that can be visually depicted, or modified. Graphics and visual representations of material, used separately or in conjunction with one another, provide a powerful medium in which knowledge or processes can be illustrated or from which concepts and judgments can be derived. Furthermore, we believe that dynamic graphics and video motion sequences allow the interactive video system to demonstrate changes, processes, and procedures as few other media can.

Interactive videodisc combines powers of the microcomputer with the image and audio storage capabilities of the optical laser disc. One side of a videodisc can hold 10 billion bits of information, which is equivalent to 54,000 video still frames or 30 minutes of motion per side. Any single frame can be retrieved in one to two seconds. This technology has the ability to overlay computer-generated text or graphics upon a video image, which gives it highlighting and explaining potential absent in ordinary television or non-interactive videodisc.

Features of interactive videodisc systems include: automatic frame recall and stop; dual audio tracks, adjustable forward, and reverse; speed control; and freeze-frame. Learners can interact with the system by touching the screen, typing answers or choices on a keyboard, using a light pen, or through voice recognition and speech synthesis. The most important feature of this technology is its ability to mix digital, analog, audio (up to 75 hours of compressed audio per side), film, and photographic presentations in any conceivable combination—a process unattainable in the past.

Some of the unique characteristics of videodisc that make it a promising medium for teaching are as follows:

1. Interactive videodisc systems provide for forms of presentation (i.e., rich visuals and audio sequences, graphics, and overlay graphics) (Bunderson, 1980).

2. Interaction is a major strength of a videodisc system. Interactive discovery or inquiry approach which requires participation and user control enhances learning, problem-solving, and decision-making skills. Learners can control their own learning sequence, content, forms of representation, speed of presentation (slow motion, fast motion, or still frames), and overall pace (Bruner, 1966; Merrill, 1979).

3. Interactive videodisc can provide immediate and appropriate feedback and reinforcement because material can be presented according to the needs and ability level of individual learners. The management and recordkeeping capabilities of the system allow for cumulative records to be kept, which encourages individualization (O'Shea and Self, 1983).

4. Because of their vast storage capacity, videodisc systems provide multiple ways of accessing information including opportunities for realistic practice, multiple examples, problems, exhibits, etc. (Bunderson, 1981).

5. The videodisc can add interest, enthusiasm, and motivation due to the intrinsic appeal of visual images, simulations, feedback, individualized instruction, and games (Malone, 1981).

Some Educational Strategies Drawn from Scanning Selected Learning Theories

Instead of beginning with the tradition of computer-assisted instruction (CAI) and projecting possible advantages of videodisc to enhance CAI, we decided to scan the broad range of learning theories for potential expanded uses of videodisc. The research literature on learning theory is quite diverse and complex, arising from several streams of modern psychology. The behavioral and neo-behavioral psychologists draw upon the work of Pavlov, Watson, Thorndike, Hull, Skinner, Gilbert, and Gagne, to name a few of the better-known researchers. Their theories focus on learning as conditioning, shaping, reinforcing, modeling, or chaining. The cognitive/developmental/gestalt psy-

chologists, on the other hand, draw upon the work of Tolman, Wertheimer, Lewin, Piaget, Ausubel, Bruner, and Landa, to name a few. Their theories tend to focus on comprehending, conceptualizing, perceiving, mapping, and achieving levels of cognitive operations. A third stream of educators can be described as humanistic/philosophic. This group of researchers and theorists draws upon the work of Maslow, Rogers, Freire, Habermas, and Mezirow, to name a few. Educators drawing from this tradition are likely to focus on the meaning of experience, communicative action, dialogue, reflection, self-knowledge, critical awareness, values, feelings, attitudes, aspirations, choices, and perspective transformations.

As a result of scanning the theorists, our conclusion is that each of these streams of learning theory is directed toward different, yet complementary objectives and that visual operations, including the features of videodisc, serve different functions for each stream of theorists. The following summary will illustrate this thesis.

Behavioral/Neo-Behavioral Tradition and Videodisc Utility

Visual imagery within the behavioral tradition is important, particularly for modelling and demonstrating. The visual environment can be viewed as a means toward shaping. Although the earlier developments of programmed instruction that formed much of the basis of CAI did not make significant use of visual images, the use of videodisc now has the potential to enhance the behavioral tradition, particularly for psychomotor learning. The use of visual images and audio sound tracks on videodisc for reinforcement also is not to be overlooked. Videodisc now makes possible very complex games and simulations that can provide learners with immediate visual feedback resulting from their psychomotor choices and operations. Flight simulators for the training of airline pilots and astronauts are examples.

With the aid of videodisc, a learner may conveniently:

a. Scan a new behavioral environment.
b. Observe and review demonstrations, procedures, or operations.
c. Visualize desired behaviors or performance standards.
d. Visualize physical sets prior to guided responses.
e. Execute procedures and operations through guided responses, games, and simulations.
f. Receive audio and visual feedback and reinforcement from performance choices and operations.

Cognitive/Developmental/Gestalt Traditions and Videodisc Utility

This tradition, although not ignoring the contributions of the behaviorists, has emphasized cognitive functioning within the "black box." Expository or reception learning is emphasized by Ausubel (1968) while Bruner, building from Piaget (1965), emphasizes discovery learning. Discovery learning can be adaptive/diagnostic or a guided inquiry process as well as a free exploratory activity. Landa (1974), on the other hand, suggests that cognitive learning requires discovery as well as the recognition and application of algorithms, rules, and heuristic devices to cases or examples. Gestalt psychologists have pointed out the important relationships between parts and wholes and the need for cognitive closure.

For reception or expository learning, visual images are typically used to: illustrate, diagram, or map; provide examples of facts, concepts, or principles; apply rules; or test comprehension.

For discovery learning, visual images have a very different function. Pictures typically are the initial means by which learners inductively explore and extrapolate for themselves the answers to their curiosity, reasoning, and testing from the particular to the general.

Videodisc systems have the potential capacity to enhance the design of both expository and discovery-learning strategies. It is possible to provide learners with immediate access to an abundant choice of visual material for both illustrative purposes or for inductive analysis. The range of visual material can include: photos of whole collections of scientific specimens; historical and anthropological evidence; maps and places; film or videotape of natural processes, disasters, and accidents; interpersonal interaction; and historical events. The engineering videodisc lesson on why the Tacoma Bridge collapsed is a good example of guided discovery learning (Fuller, Zollman, and Campbell, 1982).

With the aid of videodisc, learners may conveniently:

a. Memorize and practice recall in response to visual cues or symbols.
b. Apply concepts, principles, and procedures to visual examples.
c. Classify visual presentations according to categories.
d. Compare symbols with realistic representations for validation.
e. Identify appropriate application of rules, algorithms, and heuristic devices to cases.
f. Diagnose problem areas of performance, malfunction or illness, and propose remedies.
g. Choose hypothetical decisions or make changes in procedures, and observe results.
h. Observe parts and identify them in relationship to wholes.
i. Observe wholes and identify essential or characteristic features of parts.
j. Observe and discover alternative forms of objects or events.
k. Compare changes to objects or environments that take place over time.
l. Draw symbolic representations or concept maps from realistic images.
m. Analyze, process, aggregate, and summarize visual data.
n. Analyze examples to formulate definitions, rules, principles, and concepts.
o. Describe, explain, and interpret interactive cause and effect relationships in events.
p. Identify limitations for and constraints to observed functions or operations.

Humanistic/Philosophic Tradition and Videodisc Utility

Habermas (1971) differentiates between three generic domains of learning or learning modes: the instrumental/practical, the social/communicative, and the emancipatory/self-reflective. The humanistic/philosophic traditions of learning have focused on the latter two of these domains, while the behaviorists and the cognitive/developmental psychologists have tended to focus upon the instrumental domain. According to Habermas, it is a mistake to assume that the mode of inquiry derived from the instrumental domain is equally appropriate for all three domains. The social/communicative domain requires reflection and interpretation of one's experience related to one's culture and society, and includes inquiry into descriptive social science, history, religion, aesthetics, law, and anthropology. The emancipatory/self-reflective domain requires examination of one's identity, reifications, childhood dilemmas, habits of perception, cultural assumptions and governing rules, roles, conventions, and social expectations which dictate the way one sees, thinks, feels, and acts. Meizrow (1981) has called this process "perspective transformation" and asserts that it is the essence of adult critical awareness. Freire (1970) describes this process as conscientization, a form of social consciousness raising that liberates and leads to praxis.

Humanistic psychologists, following in the tradition of Maslow (1954) and Rogers (1969), also have emphasized reflection on emotions, attitudes, values, aspirations, self-esteem, and self-awareness through human interaction as an essential domain for learning. What all of these humanistic/philosophical theorists have in common is a focus on the importance of one's capacity to reflect on one's psychological, cultural, political, and social experience as it has meaning for the self.

Educators in the humanities, including religion, have long appreciated the importance of visual objects and events (visual arts, architecture, cultural artifacts, and symbolic acts) as the catalyst for critical awareness learning. While dialogue is the central mode for consensual validation, visual imagery can provide the focus for intrapersonal, as well as interpersonal, dialogue. Videodisc systems now enable learners to access and reflect on a broad range of visual objects and events that can provide the focus for social/communicative and emancipatory/self reflective learning.

With the aid of videodisc, the learner may conveniently:

    a. Orient oneself to new environments or cultural situations.

    b. Analyze and compare objects and events from different cultures.

    c. Appreciate and critique presentations of the visual arts.

    d. Experience vicarious symbolic action and reflect on its personal meaning.

    e. Evaluate and judge relative qualities and values in visual objects or events.

    f. Identify and question cultural assumptions underlying social behavior.

    g. Reflect on evidence of social contradictions and propose alternative actions.

    h. Validate or test hypotheses, myths, and reifications against visual data.

    i. Explore and clarify commitments to personal values.

    j. Question and analyze the validity of moral decisions.

    k. Explore and expand personal aspirations and goals.

    l. Recognize emotions in visual situations and define personal feelings.

    m. Observe visual images to stimulate creative self-expression or visions of the future.

Summary

We believe that intelligent videodisc systems provide an especially promising resource for addressing higher order learning within the cognitive/developmental/gestalt psychological tradition. Videodisc systems can provide the opportunity to present cases or problems which involve learners in the manipulation, analysis of information, synthesis, or evaluation levels, rather than merely requiring drill and practice responses at the comprehension level. In addition, we believe that the visual and audio storage and retrieval capacity of videodisc systems also can expand the instructional design and dialogue possibilities for educators within the tradition of the humanistic psychologists and humanities educators through the use of learner operations suggested in this article.

It is imperative that educators continue to evaluate the implications, assets, and limitations for the best instructional applications of this new media. We also recognize that videodisc can provide instrumentation for research, particularly on higher order cognitive learning and critical reflectivity.

A chart displaying the three learning modes, theoretical traditions, major learning theorists, associated concepts, and potential videodisc utility is presented in Figure 1.   □

References

Ausubel, D.P. *Educational Psychology: A Cognitive View.* New York: Holt, Rinehart, and Winston.

Bruner, J.S. *Toward a Theory of Instruction.* New York: Norton, 1966.

Bunderson, C.V. Instructional Strategies for Videodisc Courseware: The McGraw-Hill Disc. *Journal of Educational Technology Systems*, 1980, *8*, 3.

Bunderson, C.V. *Proof-of-Concept Demonstration and Comparative Evaluation of a Prototype Intelligent Videodisc System* (Report No. SED-790000749). Washington, DC: National Science Foundation, 1981.

Freire, P. *Pedagogy of the Oppressed.* New York: Herder and Herder, 1970.

Fuller, R., Zollman, D., and Campbell, T. Videodiscs: The Puzzle of the Tacoma Narrows Bridge Collapse Videodisc. New York: John Wiley and Sons, 1982.

Gagne, R.M. *Condition of Learning* (3rd. ed.). New York: Holt, Rinehart, and Winston, 1977.

Gilbert, T.F. Mathetics: The Technology of Education. *Journal of Mathetics, Vols. 1 and 2.* (1961). Reprinted as supplement No. 1 of the *Review of Educational Cybernetics and Applied Linguistics.* London: Longman, 1969.

Habermas, J. *Knowledge and Human Interests.* Boston: Beacon Press, 1971.

Hull, C.L. *Principles of Behavior.* New York: Appleton-Century-Crofts, 1943.

Landa, L.N. *Algorithmization in Learning and Instruction.* Englewood Cliffs, NJ: Educational Technology Publications, 1974.

Lewin, K. Field Theory of Learning. In R.B. Henry (Ed.), *The Psychology of Learning. Forty First Yearbook of the National Society for the Study of Education, Part II.* Chicago: University of Chicago Press, 1942.

Malone, T.W. Toward a Theory of Intrinsically Motivating Instruction. *Cognitive Science*, 1981, *4*, 333-369.

Maslow, A.H. *Motivation and Learning.* New York: Harper and Brothers, 1954.

Figure 1

Summary of Learning Theories and Potential Videodisc Utility

LEARNING MODES	TRADITIONS	THEORISTS	CONCEPTS	VIDEODISC UTILITY
Instrumental	Behaviorist	Pavlov Watson Thorndike Hull Skinner Gilbert	Association Conditioning Shaping Modeling Reinforcing Chaining	Demonstration Feedback Practice
	Neo-Behaviorist	Gagne	Processes Levels	Simulation
	Cognitive/ Developmental/ Gestalt	Tolman Wertheimer Lewin Piaget Ausubel Bruner	Comprehending Conceptualizing Perception Stages Exposition Discovery	Visual recall Visual examples Diagnosis Records of change Result observation Object/events-data Image extrapolation
		Landa	Algorithms	Application of rules Rules extrapolation Inferential analysis
Social/ Communicative	Humanistic/ Philosophic	Freire Habermas	Conscientization Critical-reflectivity	Cultural comparison Ethical evaluation Artistic criticism Aesthetic appreciation Role exploration
Emancipatory/ Self-Reflective		Maslow Rogers	Self-actualization Self-understanding	Creativity stimulation Emotional recognition
		Mezirow	Perspective-transformation	Values comparison Symbolic participation

Merrill, M.D. Learner Control of Conscious Cognitive Processing. Paper presented at annual meeting of the American Education Research Association, San Francisco, 1979.

Mezirow, J. A Critical Theory of Adult Learning and Education. *Adult Education*, 1981, *32*(1), 3-24.

O'Shea, T., and Self, J. *Learning and Teaching with Computers.* Englewood Cliffs, NJ: Prentice-Hall, 1983.

Pavlov, J.P. *Conditioned Reflexes* (Translated by Anrep, G.V.). London: Oxford University Press, 1927.

Piaget, J. *The Development of Thought.* Oxford: Blackwell, 1978.

Rogers, C. *Freedom to Learn.* Columbus, OH: Charles E. Merrill, 1969.

Skinner, B.F. Are Theories of Learning Necessary? *Psychological Review*, 1950, *11*, 221-233.

Saloman, G. *Interaction of Media, Cognition, and Learning.* San Francisco: Jossey-Bass, 1979.

Thorndike, E.L. *The Fundamentals of Learning.* New York: Teachers College, Columbia University, 1932.

Tolman, E.C. *Purposive Behavior in Animals and Man.* New York: Appleton-Century-Crofts, 1932.

Watson, J.B. *Behaviorism.* Chicago: University of Chicago Press, 1930.

Wertheimer, M. *Productive Thinking* (2nd Ed.). New York: Harper and Row, 1959.

Using Interactive Video for Group Instruction

William D. Milheim and Alan D. Evans

Introduction

There is a growing body of information concerning the instructional use of interactive video in education and industry. Beginning in the late 1970s, these articles have shown that this new medium can effectively combine a microcomputer and a video source (disc or tape) for instruction and/or training. This technology has had particular success in areas such as military training, health care, employee training, and special education, among others (Evans, 1986).

While subject areas like these have benefited greatly from this new technology, most such studies report the use of this medium as a learning tool in individualized instruction. However, the flexibility of interactive video also allows this medium to be used for group learning and group interaction.

Group Use of Interactive Video

Although the rapidly increasing body of research on interactive video focuses primarily on instruction for individual learners, a number of studies do indicate the effectiveness of this new medium with groups of learners.

Interactive video training at the IBM Corporate Management Development Center compared large-group, small-group, and individualized instruction with traditional lecture-based instruction. Although the individualized instruction was not completed because of time constraints, the percentage of students reaching mastery in the interactive video groups was 300 percent higher than the group learning in the lecture classroom. An infrared remote control was also found to be preferred over a touchscreen in this group learning situation since it allowed for instructor mobility in the classroom (Vadas, 1986).

William D. Milheim is an Instructional Design Specialist, and Alan D. Evans is Assistant Professor and Director of the Instructional Resources Center, Kent State University, Kent, Ohio.

Another study (Daynes, Brown, and Newman, 1981) described the effective use of Level I interactive video by elementary physical education teachers in tumbling classes. The system was well liked by teachers who learned to use the equipment instructionally after only 80 to 90 minutes of use.

Effective group use of this medium was also shown by Malouf, MacArthur, and Radin (1986), who described using interactive video (tape) to teach on-the-job social skills to learning disabled and mildly mentally retarded adolescents. Using videotape segments followed by multiple choice or discussion questions, this method proved to be more effective than an illustrated workbook with discussion and role-playing. Teachers using this interactive videotape were able to repeat all or a portion of the tape segment, continue discussion, or proceed to the next appropriate segment when ready.

Finally, interactive video was used in conjunction with hands-on training in the military to simulate operations over urbanized terrain. Used in small groups, this training proved superior to hands-on training alone, with the combination training showing significantly higher posttest scores (King and Reeves, 1986).

Rationale for Group Use

Interactive video should be considered for group instruction for a number of different reasons. Specifically, these include:

1. The fact that many students simply learn better in a group instructional setting where interaction with others can be very stimulating. The different experiences of various students can also enrich the learning situation.

2. The relatively high cost of interactive video equipment will often preclude the purchase of enough instructional units to implement effective individualized instruction. Instead of purchasing five or six separate instructional systems for individual student use, a school district need only purchase a single unit for an entire class.

3. The use of interactive video for individualized learning not only requires the large monetary expenditure described above, but also necessitates redesigning instructional space and rescheduling teacher time—changes which can frequently meet with some resistance. Since change with a group of students in a regular classroom is easier to implement (Brickell, 1961), this should be considered as a first step in adopting this new technology.

4. While most authoring systems languages previously encouraged the design and use of interactive video for individualized instruction, some

newer systems currently offer the teacher options which are very helpful for group use. One such system (LaserWorks) not only displays a pre-designed lesson, but also allows the teacher to interrupt the standard program and access other portions of the disc or repeat previous portions as desired (Blodget, 1986).

"Critical Incidents in Discipline"

Based on the rationale described above, a set of interactive video materials titled "Critical Incidents in Discipline" was designed and produced to give students, first-year teachers, and experienced teachers a sample of the discipline problems they might encounter in a classroom. Interactive video was chosen as the instructional medium since it would accurately simulate classroom incidents through the use of "real" video sequences and selectively present further sequences based on learner choice. The choices of the students using these materials would be stored by the system and later retrieved for analysis.

The interactive learning package consisted of an optical laserdisc originally mastered from one-inch videotape and a Apple microcomputer diskette authored in Apple SuperPILOT. All instructional materials were designed and produced during the spring and summer of 1986 by a team of students and educators with the help of a grant from the Martha Holden Jennings Foundation.

The hardware that connected the videodisc player and the Apple computer was an Allen VMI interfaceboard residing in slot No. 4 of the Apple IIe microcomputer. The various video cables from the disc player and the computer were routed to the monitor through a box connected to this interface. The video monitor then displayed either disc video or computer output as needed.

Use of this Interactive Video Program

The instructional materials consisted of a set of three open-ended scenarios which depicted troublesome classroom behavior. After watching each scenario, the group of learners (students and student teachers) were asked which one of a set of responses they would carry out if they were confronted with this situation in their classroom. After each choice was made, the videodisc was programmed to show what the result of that choice might be. Another scene requiring learner choice was then shown.

This sequence (presenting a video screen, requiring learner response, presenting another scene, etc.) was used in all three scenarios. For example, one scenario showed an unruly classroom with no teacher control. After being shown this scene, the learners were asked what response they might use. If they chose a somewhat neutral response, the videodisc might show the unruly class becoming even worse.

Users of this system were told, of course, that the results in a real-life classroom would certainly depend on a number of other circumstances that obviously could not be totally addressed in a single interactive video program. However, since the students were able to try a number of different choices and see the results of each, one important use of this program was the stimulation of class discussion which occurred after the students saw the results of each choice.

The Effectiveness of Group Instruction

The stimulation of thought and group discussion was therefore one of the primary instructional uses for this program. In this way it was used much the same way as a trigger film or videotape might be used (e.g., play the video sequence and then ask the students what decision they would make). However, linear films and videotapes can do this only in a very limited manner, with each progressive scene based only loosely on previous scenes or the results of student choices.

Interactive video, on the other hand, can present a series of precise video sequences specifically based on learner input from the entire group. Such input is easily obtained through a consensus of the learners who are watching the scenes. In situations where the group does not reach a consensus, several different choices can be tried with the learners seeing the results of each. Such precise control over the choice of video sequences is particularly important with complex situations or in areas where many different choices are possible, as is the case with discipline problems in the classroom.

Not only does interactive video allow learners to choose appropriate video sequences, it also allows for consequence remediation, where learners are able to see the results of particular actions. Thus, in "Critical Incidents in Discipline," not only do groups of learners get to choose what they would do, they also see the results of their chosen actions. This is in direct contrast to a simple trigger film, where learners see an opening scenario, make a choice, but then do not actually see the results of their decisions.

Another added feature of using this package with a group of learners is that the entire group has the opportunity to see the results of a variety of choices from a number of different individuals. While a single learner might not try some of the various possibilities if using the package alone, group use would help guarantee a wider variety of choices. This is particularly meaningful for inexperienced teachers looking for methods to use in their own classrooms, since someone else may give them an idea they would not have considered on their own.

Seeing others make different choices also has the added advantage of stimulating group discussion and problem-solving. If such materials were used individually, this valuable interaction between students (and with the teacher) is lost, since the student would be interacting only with the computer. As is often the case, students may learn more from unintentional classroom activities and interaction than from situations that are pre-planned.

Finally, the use of this interactive video program in comparison to linear films or videotapes allows for the gathering of data with the microcomputer that would not be possible with linear video playback equipment. In addition to learning more about how students and student teachers would handle discipline problems, this data can also be used to design and produce more effective instructional packages tailored to the needs of the students who use them.

Conclusion

While interactive video has been shown by a variety of studies to be effective in a number of instructional situations, the use of this medium in the past has been restricted largely to students learning on their own. The present article describes a method whereby this new medium can be used with groups of students to stimulate group discussion and interaction. Although such activities can often be stimulated through the use of film or videotape, interactive video may be more effective for stimulating discussion of complex scenarios or in situations where the results of learner choices should be visually displayed. □

References

Blodget, R.L. Ten Uses of Videodiscs in Public Schools. Paper presented at the National Videodisc Symposium for Education: A National Plan, Lincoln, Nebraska, November 12-14, 1986.

Brickell, H.M. *Organizing New York State for Educational Change.* State University of New York, Office of the President of the University and Commissioner of Education, 1961.

Daynes, R., Brown, R., and Newman, D.L. Field Test Evaluation of Teaching with Videodiscs. *E & ITV*, 1981, *13*(3), 54-58.

Evans, A.D. Interactive Video Research: Past Studies and Directions for Future Research. *International Journal of Instructional Media*, 1986, *13*(4), 241-248.

King, J.M., and Reeves, T.C. Evaluation of a Group-Based Interactive Videodisc System for Military Training. Paper presented at the Association for Educational Communications and Technology Annual Conference. Las Vegas, January 1986.

Malouf, D.B., MacArthur, C.A., and Radin, S. Using Interactive Videotape-Based Instruction to Teach On-the-Job Social Skills to Handicapped Adolescents. *Journal of Computer-Based Instruction*, 1986, *13*(4), 130-133.

Vadas, J.E. Interactive Videodisc for Management Training in a Classroom Environment. Paper presented at the Eighth Annual Conference in Education and Training. Society For Applied Learning Technology, Washington, D.C., August 1986.

A Videotape Template for Pretesting the Design of an Interactive Video Program

Scott V. Fedale

Introduction

The production of an interactive video program brings with it an entirely new set of challenges to someone who has previously been designing or producing basically linear programs.

First there are the requirements of designing a program to meet the needs of a variety of possible "audiences." Then there is the writing of a script for each possible "branch" of the program with the knowledge that each branch must be separate yet related to the rest of the program. Add to this the extra planning for all the different production requirements brought on by producing many "smaller" programs which must all be a part of one "larger" program.

Then introduce an element that is somewhat foreign to many video producers—the computer program which enables the user to interact with the individual components of the interactive video production.

As a final concern, add the necessity of dealing with such previously unfamiliar concerns as "tape geography," "access time," and "field dominance" and you can see how a previously confident video producer or instructional designer can suddenly feel somewhat overwhelmed.

Does Your Flowchart Really Work?

One of the most unsettling of these new challenges for many inexperienced producers of interactive video is the question of how to determine if their flowchart will actually "work" once it is transformed into a videotape or disc and a computer program.

The problem is that no amount of work with the flowchart and three-by-five cards or design worksheets adequately substitutes for the experience of actually running the interactive program and seeing what happens when the user chooses a particular

Scott V. Fedale is Associate Agricultural Editor, TV/Video/AV, Agricultural Communications Center, University of Idaho.

option in the program or answers a question a certain way.

Does the program keep the user adequately informed at all times about what is going on? Does the user get the proper feedback in relation to his or her response or progress through the program? If you are using videotape and a microcomputer, is access time a problem for any parts of the program; and, if it is, can the problem be solved by duplicating certain still frames or video sequences onto several places on your videotape? The answers to these and other questions are extremely difficult to discern from your design flowchart.

The path taken for most interactive projects is to produce all the video, still frame, and audio material, edit it onto the master video, and then make a dub of the videotape. Then add a computer code track to the videotape (if necessary) and either make a dub to use in programming the interactive videotape program, or send off the master tape to be pressed to disc. Finally, the computer program is written and the finished program is tried out with the videotape or videodisc.

In a very few instances, everything runs fine and no changes need to be made. But in most cases some changes need to be made to either the computer program, the videotape or disc, or both. Changes to the computer program are not that much of a problem, since a computer program lends itself to being entered at a variety of points, modified at these points, and then being "closed up" again after it is repaired.

However, changing the content of a videodisc means re-mastering, which is an expensive proposition. One way to save some of this expense is to use the new "check-disc" service offered by some disc mastering houses, which is considerably less expensive than pressing a new standard disc master. But you are still faced with the expense of possibly having to shoot additional video sequences or still frames, record new narration or sound effects, edit these into the program (the videotape requires that the entire program be re-edited after the point where a change is needed), and then make a new videotape dub or "check-disc." The money and time needed to make a new dub or check-disc are minor when compared with the expense involved with additional location shooting and post-production.

A Videotape Template

I recently developed what I think is a workable solution to at least part of the problem of how to test out an interactive video program before spending a lot of money on videotape production and editing. I was supervising a directed study which involved the designing, flowcharting, and computer programming for an interactive video module on nursing skills, to be used by both first- and second-year nursing students.

The student initially approached me about the directed study with the intent of producing a finished interactive video program during the semester. I informed her that since she had never done any interactive work before and since she would have to rely on a unit outside of her department to handle the needed videotape production, a more realistic goal would be to design and flowchart the project and then write the computer program, but not to get involved in any actual videotape production during the directed study. After some discussion, she agreed and we worked out a tentative schedule for the semester's work.

As we progressed through the project, I sensed a need for her to be able to experience the outcome of her efforts in a more meaningful way than simply ending up with a flowchart, a set of three-by-five cards, and a computer program. These are vital components in the production of an interactive video program and provided an excellent framework for my student's first efforts in interactive video. However, what was missing was the actual *interactive experience*: what it would be like to be a student nurse going through the program, an experience I felt was essential in order to properly evaluate and "de-bug" her instructional design and computer program.

In retrospect, my solution to the problem now seems rather simple. I constructed a videotape "template" of her flowchart, using our in-house character generator to produce "screens" which represented video sequences, menus, still frames, and graphics. We used a different background color for each screen, depending on whether it represented a menu, video sequence, etc. Thus, all video sequences had blue backgrounds, all still frames green backgrounds, etc. The color variations just served as a further reinforcement for what each screen represented in the finished production.

After we finished reproducing her 125 three-by-five cards as sequences on the videotape template, we recorded the computer time code on the videotape and logged the in and out points for each video sequence. These location times were then used in writing the computer program so that when the program was run, the "student" would see the appropriate video screen to represent the information he or she would be receiving and how it would be presented. A sample screen might be a video sequence which would run ten seconds and contain the following information:

Video: Nurse demonstrates proper method of ausculating lung sounds.
Audio: Narrator describes proper procedure and student hears what it should sound like through a stethoscope if it is being done correctly.

After running through the completed interactive video module using the video template, my student and I noticed that several changes in the design and flowchart and several additional still frames and video sequences were needed.

This merely required the creation of some new "screens" with the character generator, the editing of these new screens into the videotape template master tape, and the logging of these new screens and entry of this information into the computer program.

The expense of these procedures in terms of both time and money was minimal, especially when compared with actually having to produce new video sequences, edit a new master tape, etc. Yet it provided the interactive experience, which was necessary. It transformed the flowchart into a "living, breathing" interactive program, with a minimum of time and expense.

The interactive video module has now been run through all its possible branches a number of times with several different individuals assuming the role of the nursing student. My student feels that she is now ready to begin pre-production scripting and planning for production of graphics and still frames, on-location videotaping, and editing. She hasn't shot any on-location videotape, yet she already has a feel for how a student will progress through her program, and she has been able to effectively de-bug the program before she has gone to the expense of videotape production and editing.

Since she is planning to produce this interactive program using videotape, she has also had the benefit of being able to partially simulate the tape access time problems she will encounter in her finished program, and she has incorporated this knowledge into her revised design and flowchart.

Summary and Conclusion

In reviewing this project it became obvious to everyone involved that the addition of the videotape template was an invaluable element in the effective design of this interactive video program. The beauty of the concept is the minimal outlay of time and money required when compared with the benefits it provides in terms of evaluation of the program design and computer program.

Why not give it a try on your next, or your first, interactive video project? □

Visuals for Interactive Video: Images for a New Technology (with Some Guidelines)

Roberts A. Braden

The latest technological bandwagon is interactive videodisc . . . . Interactive computer technologies [are] media in search of designs.
—D.H. Jonassen, 1985

Many of us agree with Jonassen that interactive videodisc is the latest technological bandwagon. Why the excitement? Computer-assisted instruction (CAI) is interactive. Instructional television (ITV) has video. CAI and ITV have been around for years—decades even. What's so new or special about interactive video (IAV)? Nothing and everything. Nothing is really new. Everything is special.

The reason that everything about IAV is special is that this new technology is one of those rare and wonderful examples of what happens when technologies merge. For lack of a better term, interactive video represents a technological synthesis. Coming together are two electronic technologies (computers and television) and two design technologies (instructional design and visual design). Other technologies may be converging, too, depending upon how you categorize areas of interest (e.g., work on artificial intelligence, learning and instructional theory, perception, and sundry branches of psychology and physics have had and will continue to have major influences upon the development of IAV.)

Since these technologies *are* merging, consideration of one is consideration of all. Accordingly, before narrowing our focus to the what, why, and how of IAV visuals, we ought to take a quick glance at the new or improved features and other relevant aspects of microcomputers, television, and the systematic design of instruction as well as pertinent aspects of visual design. At the risk of being

---

Roberts A. Braden is head of the Center for Educational Media and Technology, East Texas State University, Commerce, Texas.

redundant, let us state that each of these emerging technologies has an impact both upon IAV as a composite *technology* and upon IAV as a *medium* of instruction. This distinction is important because the technology exists to support the medium. Also, for many of us, media considerations are something with which we know how to deal. The large repository of baggage that goes along with "media" is part of the total IAV equations. Some of the baggage is useful. Some is a hindrance.

Relevant Computer Features for Interactive Video Images

1. *Miniaturization.* The prefix "micro" in front of computer no longer implies anything except the size of the box they package it in. This means that the features desired in the computer element of the IAV system can probably be had in a stand-alone system.

2. *Powerful microprocessors* (with sophisticated operating systems). As the architecture of microprocessor integrated circuits increases, so does the speed of operation and the potential for access to ever larger chunks of random access memory. Nothing is more critical to the ability to rapidly handle (manipulate) digitized visual displays.

3. *Peripherals and Tools.* Graphics tools ranging from simple light pens, mouse gadgets, joy sticks, and graphics tablets to complete integrated image-capture-and-manipulation systems have revolutionized the field of graphic arts. Similar tools are available for producing displays of the printed word. Image processing and word processing are the new operational terms.

Relevant Television Features for Interactive Video Images

1. *Optical disc systems.* The laser disc, which on a single side provides 30 minutes of motion video or 54,000 randomly accessible still frames, is the item that makes possible all the high performance characteristics. With the laser disc we get rapid program execution speed, massive content resources, and room to store more visuals than even the most ambitious program designer is likely to want, need, or be able to provide.

2. *Smaller tape formats.* The emergence of 1/2-inch videotape as a suitable format for delivery of IAV programming has brought both the size and price down to acceptable levels. The small, affordable VTR is the poor folks' doorway to IAV. Although premastering of videodiscs is still being done on 3/4-inch and 1-inch equipment, the quality of 1/2-inch and now of 8mm tape has risen to the point where more and more people will soon be able to afford to participate in the production of usable video images.

3. *Frame-accurate search and editing.* Although precise editing is not as critical for IAV as it is for television programming, the technology which allows motion sequences to be accessed at any designated coded location with absolute accuracy is crucial.

4. *Special Effects.* Television producers and editors today are limited in their image making only by their own creativity. These days even small studios have chroma key, character generators, multi-function switchers, film chains, and other tools to create or manipulate electronic displays. Such technically simple things as the ability to juxtapose or superimpose images are almost taken for granted in our modern world where network news programs use these and other sophisticated techniques dozens of times in each newscast.

Relevant Systematic Instructional Design Features for Interactive Video Images

1. *Research findings.* The scholarly literature of instructional design is relatively skimpy, but useful nonetheless. The fact that there is an ever-growing body of information that provides guidelines for good practice cannot be ignored. Some of that literature is quite helpful and some is specific to IAV, e.g., see bibliographies by Kozen (1983) and Brodeur (1985).

2. *Authoring Systems.* These are computer programs that attempt to reduce computer programming to a decision-making process, with the decisions based upon design-of-instruction factors. Eventually this kind of tool will evolve to the point where it really won't be necessary to have computer programming skills in order to design IAV lessons.

3. *Graphics Tools.* Graphics tools do for the graphics designer what authoring systems do for course writers—they help the designer tell the computer to do its stuff. Some of the more powerful graphics programs come with built-in design templates that "do it all" for the designer once a data set is entered. That means that all the designer does is enter the numbers and choose a style of display. The computer organizes, calculates, does the layout, and comes up with a finished visual. Sometimes the product even looks good.

Relevant Visual Design Features for Instructional Video Images

1. *Design constraints.* Just as the sculptor is limited by the size, grain, color, and shape of the piece of marble, so too is the IAV visual designer constrained by the dimensions and nature of the video image. Sizes vary, but the usual IAV workstation will have a so-called 12-inch or 13-inch tube, which means the video picture has a 4x5

height to width ratio and is about the size of a sheet of typing paper. On that fixed, two-dimensional field the designer must create visible verbal displays, charts, diagrams, pictures, and whatever else it takes to "show" the subject. However, in all but the color TV sequences—you know, the stuff that's like what we watch on CBS and NBC—images must be created from a limited palette of colors and the end product is evanescent (once it's gone, it's gone and must be completely regenerated if one wishes to view it again.)

2. *Image detail.* The finest state of the art computer graphic equipment is capable of producing pictures of such fine detail that they seem to be of photographic quality. High density TV yields video images of similar resolution. Few IAV designers will have access to such equipment, however, and the typical IAV learning station will be equipped with a medium resolution monitor—or worse. The designer must, therefore, create images that will be acceptable on the least capable delivery system. Even for low/medium resolution the IAV designer has enough latitude, however, to choose between several levels of representation on the visual realism continuum. The actual amount of visual detail is, of course, a function of the design and artistic *effort* invested. Yes, many kinds of electronic images *are* less labor intensive to produce than older, more traditional forms of illustration. Even so, simple line drawings are apt to predominate because they offer a favorable cost to benefits solution.

3. *Competing traditions.* Electronics and design aren't exactly oil and water, but they do represent polarized perspectives of the same reality. Video has become an artistic medium of expression in addition to being a bulwark of technology. In like manner the scientific trappings of the systems approach run at odds with the freedom of expression common to artistic illustrators. Thus we find the artistic tradition of personal statement being balanced against the architectural/engineering tradition of form following function. The alternatives are compromise or chaos. We see the points of trade-off almost as philosophical points of conflict: mechanic vs artist, format vs intuition, formula vs creativity, and objectivity vs subjectivity.

4. *Ergonomics and aesthetics.* The visual design specialist must always keep in mind that his or her work is ultimately destined to be viewed by others. These viewers can be likened to the consumers of any tangible product. Their needs and desires are a high priority concern when the product is being fabricated. Viewer comfort thus becomes a primary consideration, as represented by letters that are large enough to read, color combinations that are

easy to view, and any other of several techniques that minimize discomfort. The wise designer will even go a step further by catering to viewer preferences and sensitivities.

Critical Structural Elements of Interactive Video

Just as there are four primary technologies merging to intertwine themselves in interactive video, so too are there four critical structural elements of IAV. They are:

- Interaction
- Video Options
- Pacing Options
- Strategy Options

Each critical element is affected by hardware, software, and program design factors—all variables. In turn, each of the elements has a variety of ways that it may affect the design and selection of images. With this kind of exponential compounding of variables, it is nearly impossible to conjure a single, simple set of 1-2-3 guidelines that will work for every IAV screen display design situation. Instead, it is suggested that screen designers may find it useful to use a brainstorming job aid (see Figure 1) to reduce the number of unknowns before proceeding with the design process. A separate worksheet will be required for each of the critical elements. Worksheets of this ilk are crutches, of course, but no matter. People with perfect memories need not use theory, and the rest of us will find the task simplified.

Once the critical elements are analyzed, the power and limitations of the visual designer's situation will be known. Design of IAV still-visuals can then begin with maximum assurance that the end products will be the best, most effective instructional visuals that can be expected under the circumstances. Not the best possible—just the best that can be expected under the circumstances. That leads us to the first of three "Rules-to-go-by."

RULE 1: *Don't attempt to exceed the limitations imposed by local production constraints.* Mostly that means that it's okay to stretch your limits on special occasions, but usually it is better to stay within the system constraints to avoid overload breakdowns. With IAV there is a lurking temptation to show off the technical power of the medium rather than to pursue the instructional goal with the least cost, effort, and fanfare. That leads us to the next rule-to-go-by.

RULE 2: *Visuals should complement the instructional context.* This isn't a rule for IAV only, but the easier it is for us to deliver images, the more we need to remind ourselves of this truism. Whatever we put on the screen, it should not diminish the student's enthusiasm to learn, detract

Figure 1

Worksheet for Clarifying Critical Elements in Interactive Video

CRITICAL ELEMENT: _____	
FACTORS	IMPACT ON SCREEN DESIGN
Hardware:	
Software:	
Design:	

from the sound track or other learning materials, or waste the learner's time. If, instead, the image on the screen sustains or builds interest, adds to the composite body of learning materials, and offers an efficient learning opportunity, the visual will work and the designer will have been successful.

Rules 1 and 2 deal in generalities. The first provides advice about the environment. The second speaks to our purpose—instruction. One more rule is needed that offers a prescription for dealing with the details of creating visuals. Therefore:

RULE 3: *Within reason, attempt to apply all available "standards" and guidelines.* The line between artist and technican is hard to find. It is harder yet to find undisputed evidence that any particular kind of graphic, visual technique, or display formula is best. Nothing seems to work all of the time, and things that usually work are often improved if they are interchangeably used with other successful methods. The visual design state-of-the-art is about equal to that of medicine. If we follow accepted procedures of good practice and let research help us where it can, we will be considered good practitioners. The key words are "good practice" and "research."

The field of educational technology has generated at least a ton of materials that suggest good practice in visual design. (The author moved his files recently and knows whereof he speaks on this subject.) In all of that accumulated mass of ideas, results, advice, and examples, however, there isn't a single-document item known to the author that he could recommend as a checklist/guide-sheet that would cover enough points to really get somebody started designing IAV visuals. The list that follows, though far from exhaustive and only a first cut at the task, is offered to help fill that void.

Some Guidelines for Creating Interactive Video Images

DO acquaint yourself with the relevant research literature. Or, seek out collections of principles, guidelines, and "how to" literature that have a basis in research (e.g., see Fleming and Levie, 1978; Durrett and Trezona, 1982; Pettersson *et al.*, 1984; Petterson, 1985; Dwyer, 1978, 1983; and many, many others).

DO make displays "attractive," appealing to *both* affective and cognitive objectives (Martin and Briggs, 1986).

DO show things that can be seen but are difficult to describe. There is a temptation with IAV to overuse words because text displays are so easy to create.

DO keep the sound (if there is a sound track) and the pictures on the same subject. The object is to have the senses complement each other rather than have discontinuity. Similarly, there should also be agreement between visualized text and pictorial/graphic imagery that appears in the same display (Braden and Walker, 1983; Braden, 1982).

DO make displays easy to read or view.

DO restrict lists, sets, etc., to seven items or fewer per display (Miller, 1956). Longer/larger groupings can be broken up in a variety of ways.

DO show only one primary concept or idea per visual display. (Several facts may be displayed so long as they all refer to a single, central subject.)

DO repeat whole displays or major segments of displays to reinforce learning.

DO repeat visual themes from one display to another. Repetition should be carefully planned to serve as a continuity bond as well as a memory aid.

DO use compatible color combinations. Pettersson *et al.* (1984) have identified more than 30 acceptable two-color combinations for reading from CRTs. Contrast seems to be the most important consideration—light images against dark backgrounds or dark against light. Dark backgrounds (black or brown) with white, yellow, green-yellow, grey, and cyan text tend to rank highest.

DO use an appropriate font size, e.g., use 25 characters of text per line for "TV applications," i.e., if the image will be shown on a large screen to a group (Utz, 1983). For individualized interactive "frame" formats, use 40-45 characters per line (Olson and Wilson, 1985; Grabinger and Amedeo, 1985). Occasionally use other sizes for variety or emphasis. For long segments of text (informational reading), use 70 characters per line (Pettersson, 1985).

DO use headings and labels.

DO use a combination of upper and lower case, except for emphasis or for headings (Olson and Wilson, 1985).

DO have a system for showing emphasis. For example, add emphasis via flashing or inverse text only for *rare*, extremely critical points. Use "quotation marks," underlines, and UPPER CASE whenever they will suffice, which will be most of the time. Use ☐ boxes ☐ and → arrows ← or a different color of text for mid-level emphasis.

DO be consistent in the location of program control information, that is, put help comments, directions, etc., that refer to the present screen in a window at the top of the screen. Put advance, review, and escape codes in a bottom window.

DO show abstract concepts indirectly (by showing their effects, by association, or by accepted symbols) (Alesandrini, 1985).

DO show visual analogies when concept transfer is a learning objective. (Alesandrini, 1985)

DO limit the number of colors used. Durrett and Trezona (1982) suggest four colors for beginners and no more than seven for experienced users.

DO evaluate every screen display for legibility and visibility characteristics (Whiteside and Blohm, 1984). Size of type, resolution of detail, contrast of figure-ground, and design features determine whether images are recognizable, readable, and "look okay."

DO adopt a set of style standards and then follow them. Conklin (1985) prepared a useful checklist for IAV style with items like these:

- Titles and single lines centered.
- Even spacing between information that belongs together.
- Flush left margin and ragged right margin.

- Bullets to differentiate sequence of points.
- Two spaces after a period.
- No end-of-line hyphenation.
- Numbers below 10 are written out.
- Decimals and fractions are written as numbers.
- Consistent symbol usage.

TRY TO apply accepted artistic design principles. (e.g., as recommended in Minor and Frye, 1977, or Kemp and Dayton, 1985.)

TRY TO be upbeat. A positive attitude and a fast-spaced tempo will help keep students attentive.

TRY TO be "friendly" or personal, but don't overdo use of the student's name. Instead, concentrate on ease of interaction.

TRY TO leave enough margin on all sides. (This applies for all kinds of displays and for "windows" as well as for full-screen displays.)

TRY TO use indentation and other visual cues to show organization, hierarchy, and relationships.

TRY TO have a balance of iconic and digital displays, appealing to both right and left brain hemispheres.

TRY TO apply Pettersson's (1984) guidelines when comparisons and/or discriminations of statistics are intended:

- It is easier to distinguish between lines than between areas or volumes (for charts and graphs).
- It is easier to distinguish between horizontal lines than vertical lines and lines in all other directions.
- Use columns rather than rows for the most important comparisons.
- Set columns and rows compactly and not artificially "spaced out" to fit the page.
- Bar charts are easier to read than line graphs.
- Horizontal bar charts with supporting text are the best bar chart choice.

AVOID crowded screens (leave some open space) (Olson and Wilson, 1985).

AVOID abbreviations and acronyms that are not well known or defined in the program (Olson and Wilson, 1985).

AVOID too much detail in diagrams, drawings, and illustrations (Dwyer, 1983).

AVOID animation if other techniques will work. (The labor-to-benefit ratio is out of proportion to other visual designs.)

DON'T show things symbolically unless the meaning of the symbol is clear.

DON'T show irrelevant material.

DON'T show inaccurate information.

DON'T use fancy lettering fonts.

DON'T use color alone as a cue or prompt. Make the system usable by color-blind individuals (Durrett and Trezona, 1982).

DON'T rely solely upon text to display abstract concepts.

DON'T overuse emphasis.

DON'T use a visual element as both a meaningful symbol and an emphasis cue (e.g., if arrows are used to slow flow, don't use them to point out key concepts also).

DON'T hesitate to add to this list. □

References

Alesandrini, K. The Instructional Graphics Checklist: A Look at the Design of Graphics in Courseware. In Simonson, M., and Treimer, M. (Eds.) *Proceedings of Selected Research Paper Presentations*, RTD-AECT, Anaheim, CA. Ames, Iowa: Iowa State University Instructional Resources Center, 1985.

Braden, R.A. The Outline Graphic. Paper presented at the Association for Educational Communications and Technology Annual Convention, Dallas, 1982; ERIC ED 238 413.

Braden, R.A. Visualizing the Verbal and Verbalizing the Visual. In Braden, R.A., and Walker, A.D. (Eds.) *Seeing Ourselves: Visualization in a Social Context*. Blacksburg, VA: International Visual Literacy Association, Inc., 1983.

Brodeur, D.R. Interactive Video: Fifty-one Places to Start—An Annotated Bibliography. *Educational Technology*, 1985, *25*(5), 42-47.

Conklin, J. CBT Quality Control Checklist. Paper presented at National Society for Performance and Instruction, Chicago, 1985.

Durrett, J., and Trezona, J. How to Use Color Displays Effectively: The Elements of Color Vision and Their Implications for Programmers. *Pipeline*, 1982, 7(2), 13-16.

Dwyer, F.M. *Strategies for Improving Visual Learning: A Handbook for the Effective Selection, Design, and Use of Visualized Materials*. State College, PA: Learning Services, 1978.

Dwyer, F.M. The Dilemma of Visualized Research: Lack of Practitioner Involvement and Implementation. *Educational Considerations*, 1983, *10*(2), 9-11.

Fleming, M., and Levie, W.H. *Instructional Message Design: Principles from the Behavioral Sciences*. Englewood Cliffs, NJ: Educational Technology Publications, 1978.

Grabinger, R.S., and Amedeo, D. CRT Text Layout: Prominent Layout Variables. In Simonson, M., and Treimer, M. (Eds.) *Proceedings of Selected Research Paper Presentations*, RTD-AECT, Anaheim, CA. Ames, Iowa: Iowa State University Instructional Resources Center, 1985.

Jonassen, D.H. Interactive Lesson Designs: A Taxonomy. *Educational Technology*, 1985, *25*(7), 7-17.

Kemp, J.E., and Dayton, D.K. *Planning and Producing Instructional Media* (5th ed.). New York: Harper & Row, 1985.

Kozen, N. Videodisc Bibliography. *Performance & Instruction Journal*, 1983, *22*(11), 34-35.

Martin, B.L., and Briggs, L.J. *The Affective and Cognitive Domains: Integration for Instruction and Research.* Englewood Cliffs, NJ: Educational Technology Publications, 1986.

Miller, G.A. The Magical Number Seven Plus or Minus Two: Some Limits on Our Capacity for Processing Information. *The Psychological Review, 63*(2), 1956, 81-97.

Minor, E., and Frye, H.R. *Techniques for Producing Visual Instructional Media*, (2nd ed.). New York: McGraw-Hill, 1977.

Olson, S., and Wilson, D. Designing Computer Screen Displays. *Performance & Instruction Journal*, 1985, *24*(1), 16-17.

Pettersson, R. Numeric Data Presentation in Different Formats. In Thayer, N., and Clayton, S. (Eds.) *Visual Literacy: Cruising into the Future.* Bloomington, IN: Western Sun, 1985.

Pettersson, R., Carlson, J., Isacsson, A., Kollerbaur, A., Randerz, K. Color Information Displays and Reading Efforts. CLEA Report No. 18. Stockholm, Sweden: The Royal Institute of Technology, University of Stockholm, March 1984, 14-27.

Utz, P. A Screenful of Choice Words: Using Character Generators. *Audio Visual Directions*, April 1983, 25-29.

Whiteside, C., and Blohm, P. The Effects of CRT Text Color and Decision Making. In Thayer, N., and Clayton, S. (Eds.) *Visual Literacy: Cruising into the Future.* Bloomington, IN: Western Sun, 1985.

Storyboarding for Interactive Videodisc Courseware

James F. Johnson, Kristine L. Widerquist,
Joanne Birdsell, and Albert E. Miller

Introduction

Interactive videodisc courseware systems can combine the features considered most likely to facilitate learning: student participation and individualized instruction. Yet the excitement and promise these systems engender can dissipate rapidly when the effort is made to actually develop one. The array of expertise necessary to execute such a project can be daunting, and few start-to-finish guides exist for this new technology.[1]

New integrated hardware and software systems and instructional design models, however, are making the process more manageable. Most of these current design models for videodisc development include the production of a "storyboard" as part of the design effort. But very little explanation of how this should be done and what form the storyboard should take is actually available. Consequently, the University of Notre Dame Interactive Videodisc Laboratory staff has made an effort to develop a useful method. This method will be presented, as will the forms which have been developed to document the process. A discussion of why several available forms were rejected will also be presented.

Much of the material for this article came from a cooperative teacher training and courseware development project involving a secondary school science curriculum.[2] This program involved training two high school physics teachers to participate in the design and production of level III videodisc courseware for integration into their classrooms.

James F. Johnson is Courseware Designer, Department of Chemistry; Kristine L. Widerquist is Project Manager, Interactive Videodisc Laboratory; Joanne Birdsell is Design and Production Coordinator, Interactive Videodisc Laboratory; and Albert E. Miller is Director, Interactive Videodisc Laboratory, Department of Metallurgical Engineering and Materials Science, all at the University of Notre Dame, Notre Dame, Indiana.

Three Phases of Videodisc Development

The teacher training course was structured around three phases of videodisc development: design, production, and implementation. In the design phase the content of the courseware and its presentation are determined. During the production phase everything necessary for making, programming, and delivering the courseware is done. The implementation phase consists of using the courseware in the environment for which it was intended. (The components of each phase are illustrated in Figure 1.)

The first three components of the design phase collectively provide the content required for the storyboarding process. Initially, needs are defined and these needs determine the goals of the courseware. Next, a content outline is organized, which is then used as a basis for formulating behavioral objectives.[3] These objectives will eventually be the instrument by which the effectiveness of the courseware is assessed. The behavioral objectives are organized into a learning hierarchy by successive levels of complexity. This represents the path a learner will follow through the material. Finally, the learning hierarchy is used to create a chapter hierarchy[4] which structures the objectives into lessons. Chapters involving learning capabilities that stress knowledge and comprehension[5] are designed as prerequisites to those lessons involving more abstract objectives.

Storyboarding is the final component of the design phase. At its completion, a document is provided to the production team which represents the eventual form of the courseware. Storyboarding as a term is not new. In its most basic sense, storyboarding involves the determination of the actual content the student will see. Storyboarding's application to videodisc, however, requires the assembly of information not only considered correct by content experts, but which is educationally sound, intelligible to students, and legible to production staff.

The Storyboarding Process

The Interactive Videodisc Laboratory employs a storyboarding process which involves three steps:
1. Group Storyboarding.
2. Formative Evaluation.
3. Generation of Production Documents.

During the first step a group representing all of the interests involved generates a storyboard worksheet draft containing the courseware's content. The storyboard worksheet draft is pedagogically refined during a formative evaluation. This final storyboard worksheet serves as the basis for the documents which will, in effect, become blueprints for the production staff.

Figure 1

Development of Interactive
Videodisc Instruction

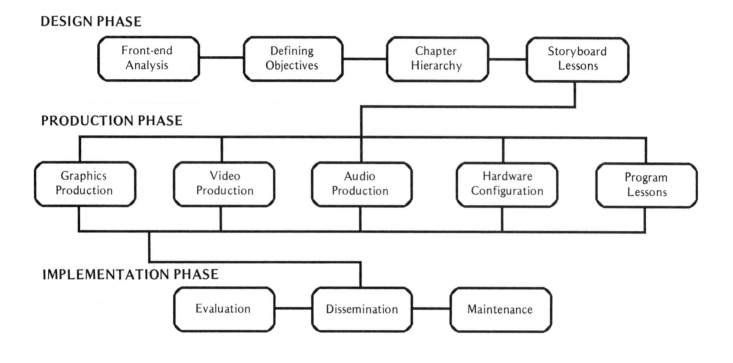

Step 1: Group Storyboarding

The generation of the storyboard worksheet draft requires a team approach. Because videodisc courseware is really a mixture of media—digital, analog, graphic, and instructional—brought to bear on some subject matter, its design requires that each of these media be represented for a credible end product. Group storyboarding acts as a funnel through which these various interests pass and blend into a whole representative of the component expertise.

The choice of team members should reflect the type of courseware being developed, the limits of the production facilities, and the hardware system from which the instruction will eventually be delivered. In general, the team should consist of these basic members: a content expert, an instructional designer, an artist or visual designer, and someone knowledgeable in production and delivery capabilities.

The team members have responsibilities corresponding to their expertise. The content expert supplies accurate information for presentation. If possible, dialogue between two content experts during drafting is very helpful. (Unforeseen gaps, however, can be pinpointed during formative evalu-

ation as well.) The instructional designer is necessary to ensure the educational soundness of the presentation and structure of the content. Because much of the power of videodisc-based instruction lies in its visual impact, it is essential to stress good visual design in the storyboarding process. Therefore, an artist or designer should be included. The production team must also be represented because the design must remain within the parameters of available production facilities. The number of people needed to fill this requirement fluctuates with the complexity of production capabilities.

Once assembled, all members of this team contribute to the storyboard draft. A cooperative group effort should be established as soon as possible. It is extremely important that this group storyboarding process remain an open, equitable dialogue. Each member, representing a certain expertise, must understand not only the importance but also the scope of his or her contribution. This understanding is fundamental to creative consensus. An optimal process of this sort will result in a whole that is greater than the sum of its parts.

The storyboard worksheet draft must be designed at both chapter and frame levels. The

Figure 2

*Sample of Initial Newsprint
Instructional Frame*

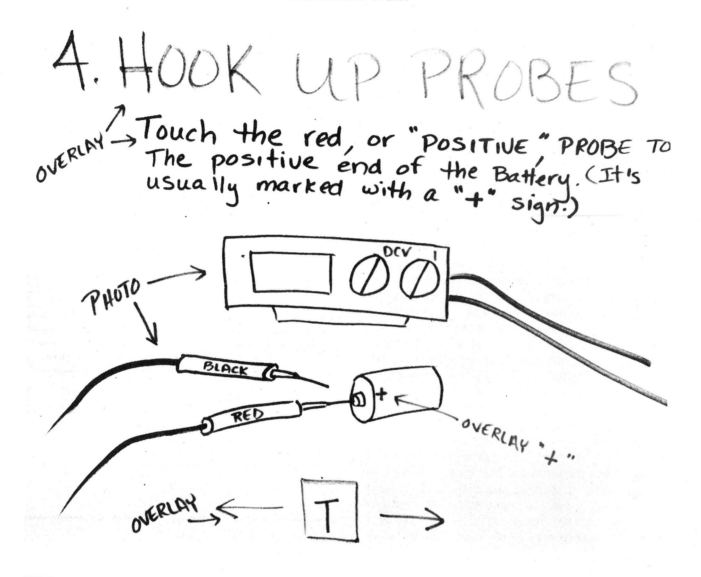

A. HOOK UP PROBES

OVERLAY → Touch the red, or "POSITIVE" PROBE TO
The positive end of the Battery. (It's
usually marked with a "+" sign.)

PHOTO →

DCV

BLACK

RED

OVERLAY "+"

OVERLAY → ← T →

previously discussed learning and chapter hier-archies provide the general organization of objec-tives within and among chapters. The team then uses a model based on events of instruction[6] to guide the sequencing of frames, and development of chapter design. The pertinent events of instruc-tion are concerned with:

(a) introducing the material;
(b) stating the objective;
(c) involving the learner via interactive dia-logue;
(d) assessing learner performance; and
(e) remediating, if necessary

Using this model the team designs the content of each frame and the flow (branching) between them.

The actual creation of a frame involves docu-menting the group input and placing the document in instructional sequence where it can be seen by the entire team. Design teams at the Interactive Videodisc Laboratory have developed a method by which the artist transfers the group consensus about each frame onto a large piece of newsprint using marker pens (Figure 2). Each newsprint frame is then physically positioned (usually on a wall) to afford ready viewing by all members of the group. At the completion of a chapter the frame contents are transferred to more manageable storyboard worksheets. It is the team's prerogative to decide whether to continue drafting each chapter as a group after establishing a pattern for the first few, or to delegate it to the content expert. If the latter is the case, drafting can be done directly onto storyboard worksheets, though they should be subjected to review by the team before any formative evaluation is done.

The storyboard worksheet (Figure 3) is the first of three forms developed at the Interactive Videodisc Laboratory for use in the storyboarding process. While seeking a document which accurately would translate the team design effort into a manageable form and also be appropriate for independent use by the content expert, the Laboratory staff reviewed several existing forms.[7],[8],[9],[10] Most of these forms are geared to providing technical and production information at the expense of displaying the actual visual content, as is most software used for storyboarding. In addition, many of the forms presupposed different production capabilities (such as level II programs) than the Laboratory had available. The instructional designer also wanted a mock-up of the actual program which could be shown to students during formative evaluation. It was determined quickly that the form would have to be simple if it were to be useful. It became clear that the different steps of storyboarding had different documentation requirements, and only when the design was evaluated and finished would it be necessary to translate it into production terms.

Unlike most available storyboard forms the Storyboard Worksheet does not contain technical or production information. This is the case for two reasons. First, it relieves designers and content experts from having to make technical decisions for which they are not qualified; and, second, it allows the worksheet to be used for formative evaluation. The storyboard worksheets can be either the documentation of the team effort or a design tool for the content expert. The only information contained on the storyboard worksheets is the visual itself, any audio script, and the branching from that frame. Overlays and motion sequences are indicated by the use of transparencies attached to the form. Thus, if so delegated, the content expert can work with a form resembling, as closely as possible, the instruction as it will be viewed by the student; he or she need not be concerned with how it will be produced.

Step 2: Formative Evaluation

Formative evaluation[11] is a process which attempts to improve courseware material by testing it in various levels of the intended instructional

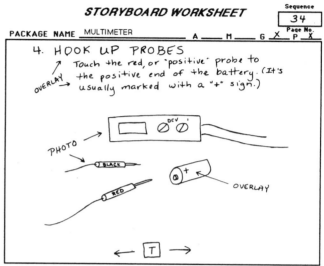

Figure 3

Storyboard Worksheet Form

environment. There are, in general, three levels of formative evaluation:

(1) individual,
(2) small group, and
(3) trial implementation.

The material is refined by sequentially passing it through each kind of formative evaluation, starting with individual evaluation and ending with trial implementation. The feedback from these evaluations aids the identification of prerequisites, student learning difficulties, gaps in the material, etc. It is not an evaluation of the method itself, or of the technology used to implement the courseware, but of the instructional design. Formative evaluation allows the detection of as many necessary changes as possible prior to the production phase. Its importance is therefore twofold; gathering information necessary for instructional revision, and minimizing the likelihood that mistakes will be carried into the production phase, after which editing becomes difficult, if not impossible.

Individual formative evaluation is a one-to-one process and has the greatest impact upon revisions of the storyboard worksheet. During this process, an evaluator from the design team, usually the content expert or the instructional designer, administers the instruction via the storyboard worksheets to a student or content expert (not of the design team). The evaluator pages through the storyboards based upon student/content expert response to the material and maintains a concurrent dialogue during this process. Comments are recorded (usually on an alternate set of storyboards) during the evaluation, and the educational effectiveness of the material is determined by comparing student results of a pre-test with an objective referenced post-test. Any mistakes in subject matter are corrected by the content expert. The output of this evaluation, composed of the evaluator's comments, student test scores, and corrections in content, is then used to update the storyboard worksheet. Usually one content expert and at least two students are exposed to the material during individual formative evaluation.

Small group evaluation follows immediately after individual evaluation, with emphasis not only on educational effectiveness but administrative feasibility as well. The updated storyboards from the individual evaluations are used as input to the small group evaluation. This evaluation can be completed with all students present as a group, or repeated serially on an individual basis. The evaluator presents the material to the student and records comments on the worksheet, but now only gives help to the student if requested. Pre- and post-tests are administered, and the student may also complete a questionnaire. A large number of students (twenty) are involved in small group evaluation; this evaluation allows the accumulation of test results (which may reveal problems not discovered during individual evaluation), learning times, and questionnaire data. These data are used collectively to further update the storyboard worksheets and produce the final storyboard draft, which will be used to construct production documents.

Trial implementation involves evaluating the courseware in actual use. While quite logical for a textbook or even a computer-assisted instructional (CAI) program, this step poses almost insurmountable problems for videodisc courseware because disc is expensive to produce and difficult to edit. It can be done, to a greater or lesser extent depending on available resources. Generally, videotape[1][2] or slide mock-ups must be used with instructor

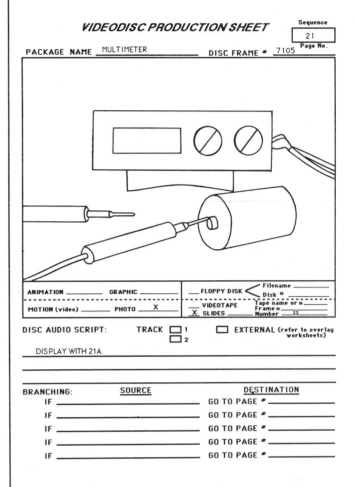

Figure 4

Videodisc Production Sheet

participation. If no reasonable facsimile can be assembled, this step must be postponed until the production phase is at least partially finished. At that time a trial implementation can and should be undertaken. While this process can still pinpoint instructional deficiencies, the data it yields may prove difficult to incorporate without major programming and production revision.

Step 3: Generation of Production Sheets

Once the revision process is complete, the final storyboard worksheets must be translated into production specifications. Two forms, the videodisc production sheet and the computer production worksheet (Figures 4 and 5), facilitate this task.

Optical laser discs can accommodate a wide range of media: still frame, graphics and photographs, animation and real motion video segments,

Figure 5

Computer Production Sheet

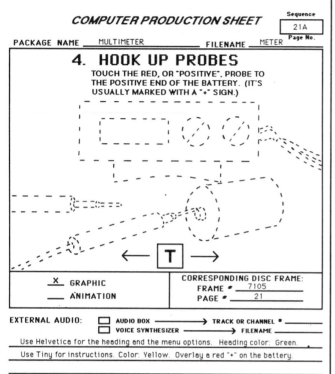

COMPUTER PRODUCTION SHEET

Sequence: 21A

PACKAGE NAME ___MULTIMETER___   FILENAME ___METER   Page No.

4. HOOK UP PROBES

TOUCH THE RED, OR "POSITIVE", PROBE TO
THE POSITIVE END OF THE BATTERY. (IT'S
USUALLY MARKED WITH A "+" SIGN.)

← T →

__X__ GRAPHIC

___ ANIMATION

CORRESPONDING DISC FRAME:
FRAME # ___7105___
PAGE # ___21___

EXTERNAL AUDIO: ☐ AUDIO BOX ——→ TRACK OR CHANNEL # ___
  ☐ VOICE SYNTHESIZER ——→ FILENAME ___

Use Helvetica for the heading and the menu options. Heading color: Green.

Use Tiny for instructions. Color: Yellow. Overlay a red "+" on the battery.

BRANCHING:	SOURCE		DESTINATION	
IF	←		GO TO PAGE #	20B
IF	→		GO TO PAGE #	20A
IF	T		GO TO PAGE #	9
IF			GO TO PAGE #	
IF			GO TO PAGE #	

audio incorporated with motion sequences or encoded onto still-frames, and even digital information. The videodisc production sheet is designed to specify each disc frame in the program. For each frame the illustration is either drawn or indicated by a written description of the video to be shot, photograph to be taken, or animation to be drawn. The type of frame or frames and their sources are specified. An audio script and the branching from that frame or segment are also included. No space is delegated for the digital code required for level II programs, as the external computer control characteristic of level III systems makes its inclusion on the disc unnecessary. Every graphic, photo, animation sequence, and motion video segment is detailed on the videodisc production sheets as the final storyboard worksheet dictates. These sheets are arranged in the order that the material will be placed on the disc.

In level III instructional programs the videodisc is, literally, only half of the picture. The external computer can, with the proper hardware, provide additional presentation capabilities such as computer-generated text and graphics, as well as non-videodisc audio sources. Other peripheral devices can also be incorporated, such as touch sensitive screens, which must be considered in the storyboarding process. Computer control makes it possible to use the disc as a video database over which instruction and information are generated by the computer in an easily edited form. Therefore, a disc frame can be used again and again to communicate different concepts by changing what the computer presents to the student. These computer presentations are detailed on computer production sheets.

Like the videodisc production sheets, the computer production sheet is largely taken up with space for a representation of the visual. Usually the corresponding disc frame is indicated by dotted lines and the computer generated image is drawn over it to provide for correct placement. The nature of the image is specified as is the corresponding disc frame, if applicable. Any external audio to be used in conjunction with this disc frame is scripted and its source noted. Finally, the branching from this composite of computer information and videodisc frame is detailed.

After the videodisc and computer production sheets for the entire course are complete, the branching information contained on these is combined into a flowchart for the program which will run the disc. The design phase is then finished. These documents are turned over to the production team and form the basis for the final courseware (Figure 6).

Conclusion

Storyboarding is the pivotal portion of the development of interactive instruction. All other design tasks flow into it and production cannot begin without it. The three steps of the storyboarding process: group storyboarding, formative evaluation, and generation of production documents, are similarly interrelated; the product of each step becoming the input of the next. This relationship is illustrated in Figure 7.

It has not been the aim of this article to discuss the principles of interactivity, or how to build them into the storyboard, nor has it been to discuss the mechanics of screen design—although these are also vitally important to successful storyboarding. The purpose has been to present educationally sound guidelines for moving from the identification of an instructional need war-

Figure 6

Photo of Final Instructional Frame

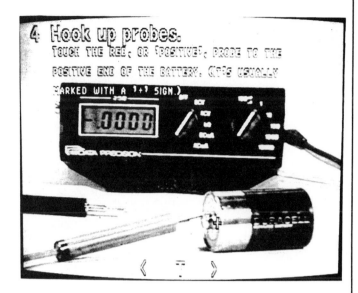

ranting videodisc courseware to actual production. Certainly this process will become more and more sophisticated with practice, and the intensive time and labor will be reduced. Yet until the idea of interactivity and videodisc technology itself becomes more familiar, these guidelines provide for dialogue between all concerned parties, allowing the final product to reflect the best of several areas of expertise. ☐

Notes

1. 3M Optical Recording Project, *Producing Interactive Videodiscs.* 3M Co., St. Paul, MN, 1982.
2. Lilly Endowment Inc. of Indiana Linkage Grant, "Secondary School Faculty Training in the Development and Use of Interactive Videodisc Courseware," 1985-1987.
3. Mager, R.F. *Preparing Instructional Objectives.* Palo Alto: Fearon Publishers, 1962.
4. Johnson, J.F. Chemical Applications for Micros. Paper presented at the Statewide Academic Microcomputer Conference, Indianapolis, April 1985.
5. Bloom, B.S. (Ed.) *Taxonomy of Educational Objec-*

tives, Handbook of the Cognitive Domain. New York: Longman: 1956.
6. Briggs, L.J. (Ed.) *Instructional Design.* Englewood Cliffs: Educational Technology Publications, 1977; Chapter 7.
7. DeBloois, M.L. (Ed.) *Videodisc/Microcomputer Courseware Design.* Englewood Cliffs: Educational Technology Publications, 1982; Chapter 2.

Figure 7

The Storyboarding Process

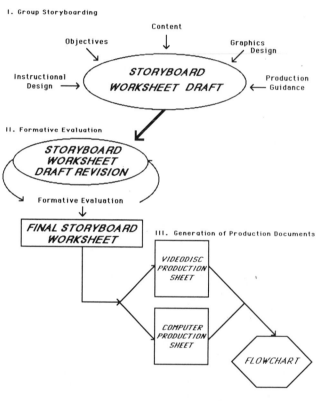

8. Smith, R.C. From Script to Screen by Computer for Interactive Videodiscs. *EITV,* January 1985, 31-33.
9. Harlow, R.B. A Personal Computer Program for Storyboarding. *EITV,* January 1985, 34-36.
10. 3M, 22-23.
11. Briggs, Chapter 10.
12. Fedale, S.V. A Videotape Template for Pretesting the Design of an Interactive Video Program. *Educational Technology,* August 1985, 25(8), 30-31.

Videodisc Simulation: Training for the Future

Jon I. Young and Paul L. Schlieve

Videodisc technology may well revolutionize education in both public and private institutions by the end of the decade. This technology has the potential to provide the most complete and effective training yet developed. When a videodisc player is connected to a microcomputer, it becomes very flexible and capable of highly sophisticated instructional sequences. For the first time, the delivery system for truly individualized instruction is available.

Videodisc application has so far been generally associated with information storage and retrieval. With up to 54,000 visuals available, consumer catalogs can be recorded on videodisc, and the built-in microprocessor can locate pictures and verbal descriptions of any product from an index. Or, if the buyer prefers to browse, the disc information can be played at regular, slow, or fast speed. Unlike videotape, the images are not distorted, but remain clear at all but the fastest speeds.

Once the consumer finds an interesting item, there can be a demonstration of how the item is used, assembled, or repaired. This can also be done with motion sequences at regular, slow, or fast speed.

Another current application is personalized travel tours. Not only can recreational industries use the system to preview potential vacation high spots, but people who are financially or physically unable to travel beyond their home can take the adventure in absentia. Through specific programming, a homebound or otherwise limited individual can visit exotic locations, exciting cities, or high adventure almost as if he or she were driving or walking himself or herself.

Instructional Application

The most significant application, however, will be in the area of instruction and training. For many years, efforts at developing instructional/training materials that account for human differences in learning, and evaluation materials that are able to certify learner competency have been produced at high cost.

Computer-controlled simulators have been used in training airplane and helicopter pilots for civilian and military aviation. These simulators are multimillion dollar investments. Videodisc simulator technology will provide quality simulation training and competency assessment at an ever-decreasing cost.

Because of the newness of this technological application, there is very little reported information available on its instructional value. However, two instructional efforts show programs which document the effectiveness of the approach.

In a report to the United States Training and Doctrine Command, the Signal Corps demonstrated that communications operators could be taught to operate complex pieces of equipment without touching the equipment until after the learning process. The videodisc simulator they developed allowed the learner to make equipment adjustments, connect cables, and read dials through a series of discrete visuals. The videodisc can show a series of discrete slides as if they were motion much like an animated movie. Learners could turn a dial showing the presentation and actually make simulated adjustments to the communications equipment.

Periodically in the program, the learner was asked to demonstrate understanding of the principles taught exactly as he or she would with the specific communications equipment. In this way, each learner could be certified as a competent machine operator before even working on the actual equipment.

A statistical comparison of learners certified using a videodisc simulator and learners certified by human trainers using the actual equipment showed no difference in actual ability to operate the complicated communications equipment (Young and Tosti, 1980).

Another major demonstration of videodisc training capability was developed for the American Heart Association. David Hon prepared a self-contained system, including a mannequin, that effectively teaches cardiopulmonary resuscitation in less time than the traditional human instructor approach. Again, the training program will certify learners as competent or in need of additional training (Hon, 1982).

These two programs have demonstrated that videodisc simulation training is as effective as traditional approaches, and, by removing human variability, this technique will provide more consistent training and assessment. No one will be able to "pass" the course unless he or she demonstrates his or her own ability and competence.

Jon I. Young is Associate Professor and Paul L. Schlieve is Assistant Professor, North Texas State University, Denton, Texas.

Program Development

In order for any simulation to be effective, it must be developed correctly. Videodisc simulation is no exception. The program will be only as accurate as the information comprising it.

As in any programmed instruction, videodisc training is developed by working backwards from competency statements. Assessment questions are created to measure each competency. A task analysis details the required information, and a narrative script with appropriate visuals is prepared.

Remedial information, alternate instructional paths, and feedback loops are carefully planned to accommodate a variety of learners. This all takes time—often many months. Since each videodisc simulation requires such a detailed analysis, development programs will not likely be developed for tasks that are easily taught some other way.

The variables that will most influence whether or not a videodisc simulation is the correct approach are incorporated in the following questions:

1. Is the training dangerous to the learner?
2. Is the training expensive either in materials or time?
3. Is consistency of training or assessment critical?
4. Is it expensive to conduct training in several locations?
5. Is an accurate competency measure critical?
6. Is the training highly complicated?
7. Is remedial or review capability important?
8. Is a qualified instructor available?
9. Is training a constantly reoccurring process?
10. Is immediate feedback to the learner critical?
11. Is a range of different presentation strategies required?
12. Are diagnostic data important?
13. Is learner accessibility important?
14. Is individualization/self-pacing important?
15. Is active learner participation important?

Answering "yes" to these questions would indicate that a videodisc simulator is a viable alternative to traditional instruction and training.

Learner Motivation

Probably the most difficult challenge of education and training is to motivate the learner to acquire the information. Motivation is highly correlated with involvement and commitment.

In this training approach, the learner is required to interact with the material he or she studies. The greater the interactivity, the more control the learner feels, and the greater the sense of commitment or motivation.

User Interactivity

The last and perhaps most impressive characteristic of the videodisc simulator is the learner's opportunity to become an active, rather than a passive, participant. Unlike traditional training programs, the videodisc system requires the learner to participate through answering questions, selecting optional material, or simply communicating with the computer.

The developer of videodisc simulation has several learner interaction modes available. Learner interaction can occur on a standard computer keyboard. This approach is most effective with sophisticated learners who are accustomed to using computers or word processing equipment.

A second mode is touch pads which operate much like television remote controls. The learner can give commands, follow directions, or even answer questions.

A third mode involves a touch-sensitive screen that allows the learner to interact by touching various sections of the monitor screen. A similar technique is to use a light pen. With this mode, a connected light pen interacts with the video screen to give directions or answer questions.

A fourth mode is illustrated by the Heart Association's mannequin techniques. Electronic sensors in the mannequin allow the computer to monitor learner application of external heart massage and artificial breathing techniques. This same principle can be used whenever the competencies require physical manipulation of some external object.

Summary

In spite of the cost of time and expense to develop videodisc simulation, the technology is a permanent part of our educational and training repertoire. As new advances and demands reduce the cost, application will extend to subjects taught in public and private school classrooms. No discipline, from science to language arts, will be exempt, and all will benefit. □

References

Hon, D. Interactive Training in Cardiopulmonary Resuscitation. *Byte*, June 1982.

Young, J.I., and Tosti, D.T. *The Effectiveness of Interactive Videodisc in Training.* U.S. Army Technical Report, 1980.

Design Considerations for Interactive Videodisc Simulations: One Case Study

Stephen R. Rodriguez

Introduction

Increasing use of interactive videodisc (IVD) as an instructional tool has recently sparked interest in the "how-to's" of IVD lesson design. Simulation—one possible style of interactive presentation—has emerged as a viable instructional format for IVD. This article presents design considerations for interactive videodisc simulations and describes some of the attributes and advantages of this format.

Videodisc Simulations: Some Background

A simulation may be defined as an activity that incorporates a model of natural or man-made phenomena and allows the learner to interact in a limited version of reality (Florini and Keller, 1981). As noted by Ruben and Lederman (1982), simulations often possess some or all of following characteristics:

(1) The learner is given (or selects) a role and may be allowed to set certain simulation parameters.
(2) Interaction occurs in that the learner manipulates simulation variables and experiences the consequences of actions or decisions.
(3) Rules govern the interactions.
(4) Interactions occur in pursuit of goals.
(5) Criteria are included for determining that the goal has been attained or that the simulation has ended.

A level III IVD system's presentation capabilities are conducive to simulation development. Those include:

(1) Storage of 30 minutes of video or of 54,000 still images on each side of the disc; both of these display capabilities may be employed on a given side;
(2) Ability to precisely access and display any portion of the playing side of the disc in three seconds or less;
(3) Storage and random accessing of two independent audio tracks that can provide a total of 60 minutes of audio per side;
(4) Ability to randomly access and display computer generated text and graphics.

Rapid accessing of video imagery and audio messages under program control enables the designer to realistically depict conditions and consequences of some performances that would otherwise be dangerous or impractical to create or practice (e.g., treating a patient, mixing volatile chemicals, or applying for a job). Often, IVD simulations compress time and include random events. They may also record and report the learner's score or progress toward a goal as well as the time expended toward attainment of that goal.

Staff of The Center for Instructional Development and Services at Florida State University recently developed a seven disc IVD series for Florida's Job Training Partnership Act (JTPA) Program. The goal of the three-year, one and a quarter million dollar project was to provide JTPA clients—"hard to employ" individuals—with the skills and knowledge that would enable them to obtain and keep jobs. The discussion of design considerations for IVD simulations that follows stems from the author's involvement as a writer and instructional designer in the project.

The Design Process

The JTPA project design team developed some five IVD simulations: employability skills addressed in these simulations included conducting a job search, applying for a job, completing a job interview, applying generic job skills at a new job, and communicating with customers, co-workers, and supervisors. Each simulation served as a cumulative practice opportunity for content covered in lessons presented earlier in a given disc.

Prior to initiating an IVD simulation design effort, the goals and objectives of the instruction and the nature of the learner were assessed to determine the viability and usefulness of developing a simulation. By classifying each objective as a particular type of learning outcome, the design team was able to create simulations that engaged the learners in activities suitable for promoting acquisition of the given skills and knowledge. Possible motivational effects were also considered. Since simulations give learners opportunities to apply new skills in meaningful contexts, simulations may increase learner motivation and shape certain affective outcomes (Keller & Suzuki, 1987). Other general questions that may be useful

Stephen R. Rodriguez is Project Manager, Center for Instructional Development and Services, College of Education, Florida State University, Tallahassee, Florida.

when considering the suitability of IVD simulation include the following:

(1) Does the subject matter lend itself to a visual treatment?

(2) Are the skills in question critical?

(3) Is practicing of the skills in the real world problematic? How important are the conditions and consequences of performance?

(4) Is learner motivation a concern? Will providing learners with the opportunity to apply newly acquired skills foster positive attitudes toward the skills and their subsequent application in life?

(5) Is a video, picture-based presentation more likely to promote learner engagement and practice than a text-based presentation?

Positive answers to most of the above questions suggested the feasibility of developing an IVD simulation.

The design team, consisting of an instructional designer, a producer, an operations manager, a programmer, and a writer/instructional designer, next grappled with some basic questions regarding the structure of the simulation. The domain of skills to be addressed in the simulation had to be identified. This step was completed collaboratively by team members during a treatment brainstorming session. The writer then developed the treatment, which was reviewed and then revised. Once finalized, the treatment served to guide scripting, storyboarding, and flowcharting of the simulation. Laugen and Hull (1987) provide an expanded description of the team approach to IVD development employed for the JTPA series.

Duke (1981) suggests that a simulation consists of some twelve basic elements. His view, however, applies more to "person-based" simulations than to those designed for IVD. Our experience with the JTPA project suggests that the basic issues in IVD simulation design include the following:

(1) **Determine the scenario:** In what context(s) will the simulation occur? What is the goal that the learner is expected to achieve? What types of characters and events are to be portrayed in order to establish a basis for presenting cues and conditions to initiate performance and to depict the consequences of performance?

(2) **Determine the perspective to be employed:** Will the learner observe an on-camera character (third person perspective) or will the camera, in effect, become the character (second person perspective)? In the first case, the learner observes one or more characters while in the second case, on-screen characters directly address the learner.

(3) **Determine how interaction will occur:** By what method will the learner be able to control or influence the course of events? What is the nature of the interactivity within the body of the simulation?

(4) **Determine how the learner's performance will be reinforced and evaluated:** What imagery and narrative content will be employed within the simulation to indicate the relative appropriateness of the learner's actions? How will the learner's overall performance be evaluated at the end of the simulation in order to indicate the extent to which the goal was attained?

(5) **Determine how to present feedback:** How will the learner receive feedback (remediation) upon completion of the simulation regarding the relative strength of choices made within the body of the simulation?

These were the major issues faced by the design team. The resulting IVD simulations thus tended to include the following major components:

(1) An introduction that established the scenario, the major character (if any), and the goal. This part also included any further directions to the learner that might be necessary. Examples of goals in JTPA simulations include making a favorable impression at a job interview, succeeding at a new job and getting a permanent position, or being an effective communicator.

(2) The body of the simulation in which the imagery used to cue or prompt learner actions, any embedded questions, and imagery used to depict the consequences of performance were presented. Interaction is of course a crucial concern here.

(3) An evaluative phase in which the learner's relative success toward attaining the goal is presented. Preferably, this evaluation consists of more than simple score reporting and is presented in a manner that is intrinsic to the narrative content of the simulation. For instance, in the simulation on good work habits, the evaluation is presented to the main character by his supervisor in the form of an actual job evaluation.

(4) A remediation phase in which the learner's decisions during the course of the simulation are reviewed and remedial or reinforcing feedback is provided as required. This part typically utilized the original video and the secondary audio track to trace the learner's choices and to review the relative strength of those choices. Of course, the learner's responses were recorded as they were made by the software.

While the simulations developed for JTPA did tend to employ the structure noted above, certain design choices had to be made for each simulation. The introductions usually consisted of brief video segments with voice-over narration. Each introduction served to establish characters, the setting, and the goal.

The bodies of most of the simulations were built upon an embedded, multiple-choice question framework. The learner saw some video and was then presented with a question via computer text as to possible responses or actions. A video segment was then accessed on the basis of the learner's choice. This cycle was repeated throughout the body of the simulation.

In addition to the embedded question style of interaction, error identification was also employed. *Good Moves*, for instance, engaged the learner in identification of errors made by an on-screen character in order to guide that character's actions in a job setting. Those errors that the learner identified were immediately corrected by the on-screen character. If an error was not identified, possible negative consequences of committing such an error were revealed. In addition to answering multiple choice questions and identifying errors, learners may interact in simulations by entering variable values, by answering a series of questions before a branch is made, or by manipulating a job-specific input device such as the "welding torch" or manikin incorporated in some of David Hon's IVD work.

The evaluation phase varied from one JTPA IVD simulation to another. In some instances, the learner's score (i.e., percentage or number of strong vs. weak choices) was presented. More commonly, evaluation was presented via the response of a character within the narrative structure of the simulation.

The final phase of each simulation consisted of remediation. This occurred after the body and evaluative phases of the simulation, thus allowing the learner to complete the simulation without obtrusive interruptions. The learner's choices during the course of the simulation were revisited by reaccessing the appropriate video segments and through display of computer-text messages. Through voice-over narration on the secondary audio channel, explanations specific to each response were provided. The learner thus received additional feedback and learning guidance and maximum use of the video material was made.

Conclusion

IVD simulations are an exciting instructional format. And, while field test results of the present project reveal that these simulations are liked by learners, many important questions related to the design and use of simulations in learning contexts remain unanswered. Simulation design issues should continue to be explored and related theories articulated and tested. Research as to the most appropriate uses of simulations and as to their instructional and motivational effects should be conducted. Design issues pertaining to so-called "adaptive simulations" (Breuer and Hajovy, 1987) should also be explored. Work in these areas may lead to refinement of useful design schemes and principles for effective IVD simulations. □

References

Breuer, K., and Hajovy, H. Adaptive Instructional Simulations to Improve Learning of Cognitive Strategies. *Educational Technology*, May 1987, *27*(5), 29-32.

Duke, R.D. A Paradigm for Game Design. In C.S. Greenblat and R.D. Duke (Eds.), *Principles and Practices of Gaming-Simulation*. Beverly Hills, CA: Sage, 1981.

Keller, J.M., and Florini, B.M. *Research on the Motivational Properties of Games and Simulations*. Unpublished manuscript, Syracuse University, Instructional Design, Development, and Evaluation, 1981.

Keller, J.M., and Suzuki, K. Use of the ARCS model in Courseware Design. In D.H. Jonassen (Ed.), *Instructional Designs for Computer Courseware*. Hillsdale, NJ: Lawrence Erlbaum Associates, 1987.

Laugen, R., and Hull, S. *Developing Videodiscs for Employability Skills Training*. Paper presented at a meeting of the Society for Applied Learning Technology, Orlando, Florida, February 1987.

Ruben, B.D., and Lederman, L.C. Instructional Simulation Gaming. *Simulation & Games*, 1982, *13*, 233-244.

A Model for the Design of a Videodisc

Joanne C. Strohmer

Designing a videodisc is a very complex process. It involves all the steps which contribute to effective video plus all the considerations necessary to create computer software. Combining the two presents some unique possibilities and problems.

It is impossible to take advantage of the tremendous capability of these two media and also keep track of the numerous details without a carefully planned system. The design model described here proved very effective in meeting both needs during the development of the *Business Disc*.

Nature of the Project

The *Business Disc* is a complex videodisc simulation. It is designed to give high school, college, and other learners a chance to use the technology to plan and try out a small business. The program, which can take users anywhere from several hours to days to complete, provides information and experiences with organizational structure, business location, pricing, computing employee expenses, and projections of cash flow.

The presentation involves motion video, stills with voice-over, Chyron graphics, and computer text and graphics. It includes special functions such as a glossary, a note pad, and printout capability.

The Project Team

Maryland Instructional Television (MITV) and Maryland Public Television (MPT) collaborated as Maryland Interactive Technologies to create this disc. MITV produced the disc, and MPT is handling distribution.

Meetings were held to discuss the general content. At that time, MITV was producing *Open for Business*, a linear series on entrepreneurship. It was decided to use the same topic for the disc. The disc and the series were designed to be compatible but not dependent on one another.

Joanne C. Strohmer, formerly with Maryland Instructional Television, is now Supervisor of Reading and Language Arts, Carroll County Public Schools, Westminster,

A producer and instructional designer were pulled out of the normal structure of MITV to form a special project team. Michael Sullivan, who at that time headed the organization, gave producer A. Ademola Ekulona and designer Joanne C. Strohmer the mission of trying to produce a level-three disc in one year. The objective of the experiment was to see if this agency, known for quality linear video, could also succeed with videodisc software.

The producer and designer had extensive backgrounds in instructional video but neither had created a level-three disc. Ekulona had taken a week-long Nebraska Videodisc workshop. Both had completed a six-week course in computer programming so they had rudimentary understanding of computer capabilities.

After the planning stages were finished, a director, Natalie Seltz, and technical experts became part of the project. None of these had previous experience with videodisc production. Since no one on staff had adequate programming expertise, an outside consultant was hired to create the computer software based on detailed specifications provided by the team.

Project Stages

The creation of the videodisc simulation fell into five overlapping stages. These included overall design, content specification, preparation for production, production, and distribution activities. The first three stages, the planning steps, took approximately four months. Video production took another four months. Four additional months were allocated for computer programming, but this activity actually took a full year.

Stage One

The first stage focused on overall design. A simulation format was chosen because it provided a chance for active involvement and application of knowledge. It allowed users to learn by doing and gave them a chance to tailor the experience to their own interests and needs. It was also felt that using this difficult format would challenge the team and provide a lot of information about the agency's capabilities to produce a disc.

With the topic and format decided, Ekulona and Strohmer began detailed planning. To get an overview of the content, the producer and designer attended workshops on starting a small business.

The flowchart and prescript package were then created. The designer and producer worked face to face, with the former sketching out flowchart blocks and the latter simultaneously writing prescript forms. Each flowchart block and its corresponding prescript form was assigned a number which was used throughout the project.

Prescript forms specified the basic nature of the information and whether it would be conveyed by video (motion, still, special effects, chyron text, chyron graphics) or computer (text, graphics) and the specific requirements within each of these categories, for example which audio channel to use. Branching information was also included.

Stage Two

Stage Two involved researching and writing content specifications. Because the flowcharting and prescripting had been so detailed, this stage could be handled very efficiently. Basically, it involved searching out the information designated on the flowchart. For instance, one block called for information on advantages and disadvantages of sole proprietorships. The task is simply finding this information and recording it.

Content was merely a very clear and exact outline of information in a looseleaf notebook. Each content page was coded with the number corresponding to its flowchart block containing exactly the points that were to be made. There was no need to put extra energy into creating an elaborate content document.

In addition to many hours in the library, content collection also involved contacting attorneys, insurance agents, and other experts to get specialized information. Once the content specifications were completed, they were reviewed for accuracy by a representative of the U.S. Small Business Administration.

Stage Three

The main activity in this stage was preparing for production. Scripting began as soon as a good number of the content specifications were written. The original plan was to have a scriptwriter handle this task entirely, but with approximately one hundred fifty segment scripts ranging from a few seconds to a couple of minutes to be generated, the producer also became involved in writing. The designer reacted to and edited scripts.

Only two drafts were generally prepared. This was possible since the content had been so exacting and because this team was operating outside of the agency's normal system of routing and responding. The scripts were, however, reviewed by content experts.

At this point, with scripts accumulating, it became obvious that a system for managing them was necessary. A wall chart was made to track the progress of each script. Again, every script was coded with its flowchart block number.

Also, with scripts being completed, set design

and casting could begin. Character sketches were written for each major role. The most challenging part to cast was the accountant, who was also the host. Since the simulation was done point-of-view, the host had to be capable of talking to the user face to face in an individualized instruction setting. The designer participated in decisions about major actors.

Stage Four

Production of motion video and stills took place during stage four. The designer reviewed all edited segments and worked with the producer on the design of Chyron graphics.

Most design activities at this time, however, related to the computer segments. Around three hundred screens were designed. Exact specifications for what was to appear and how it should be displayed were recorded on screen design forms. Preliminary decisions were also made about special functions and special function keys.

Another component created at this time was the series of "if-then" conditions to guide branching and mathematical formulas for figuring elements of the instruction, such as sales receipts, productivity, and the effects of good and bad events on the business. This document contained over eighty pages.

Paper proofs were undertaken next. These consisted of approximating use of the videodisc to discover flaws in logic or clarity. The designer guided individual volunteers through the experience, describing content relayed in the video, asking for responses based on the computer design sketches, and computing the formula with a calculator.

Each proof took an entire day but yielded much useful information about changes needed. The designer used the results of each session to make revisions and then retest with another volunteer. This process was continued until problems were no longer being found.

With this step done, the flowchart, screen designs, and formulas were turned over to the programming consultants. During this time, the designer's tasks consisted of numerous meetings and telephone calls with the programmer and reviewing work as it was done. Print materials were also begun.

Although the planning and video production remained on schedule, work bogged down with the computer programming. This seemed to be due to at least two factors. The programming required was much more complex than realized in initial meetings with the programmers. Also, having to work with programmers who had other demands on their time complicated and slowed down the effort. In fact, this problem was so critical that eventually the work was re-contracted to another company.

As pieces of the programming became available, volunteers were recruited to test them to see if directions were clear, branches functioned properly, and where bugs were encountered. An additional person was hired for several weeks to schedule these tests and record problems that were discovered.

Stage Five

As the programmers continued their work, the Maryland Public Television representative, Ralph France, who was to handle distribution, became more and more involved. In time, he took over the arduous tasks of field testing, communication with the programmers, and completion of the written documentation. This phase took many months and overlapped with the initial distribution of the disc. At this point, the producer and designer had moved on to other projects.

Implications for Other Videodisc Projects

While every interactive disc has its own unique set of challenges, some conclusions can be drawn from this project and used by designers of other discs.

Conditions to Avoid

The main shortcoming of this project was not having one or more computer programmers on staff full time. While the team had enough background to design the computer components, having a programmer involved earlier in the project might have enabled everyone to have a clearer grasp of the difficulty of what was being asked and build in time and systems for managing this aspect.

Having a programmer on staff would also have meant that the project could receive full-time attention. This would cut down the extremely long time for programming that was experienced in this project.

It would also have been helpful to spend several planning sessions at the very beginning to come up with a system for managing details such as keeping track of the scripts. This was done once the need was realized but by that time some degree of stress and confusion had occurred.

Aspects to Duplicate

The most crucial aspect of a videodisc project, whether this complex or simpler, is the planning time. Content collection, scripting, and video production moved along rapidly with few false starts, and little or no need to re-do anything because all aspects had been spelled out in such detail at the start.

A detailed flowchart is essential from the beginning. A system for handling the numerous pieces from scripts to computer designs to Chyron graphics specifications also is necessary.

Having a small team that was given the authority to work somewhat independently made it easier to keep the design and video production on schedule and under control. This kind of project requires a high degree of concentration and continuity.

People cannot move in and out of working on it with any hopes of understanding how the innumerable pieces fit together. Nor can the project be accomplished with any degree of efficiency if the team assigned to it must divide their attention between a disc project and other responsibilities. Too many details must be carried in the head and fit together. Distractions are sure to lead to mistakes.

It was important to designate one person, in this case the designer, to be the keeper of the most recent copies of the flowchart, scripts, and other related documents. Anyone who is confused about what is most current can check with this person. In order to know what is most current, everything should be dated.

Another aspect of being the keeper of the documents is serving as a "clearing house" for any changes on them. Any tiny revision in a project such as this can have a ripple effect. The designer is the person most likely to have an overview of the entire project and be able to record any changes and make the resulting adjustments throughout the program.

For instance, in this case one figure in the formula for the effect of advertising was changed during the course of the paper proofs. This caused not only changes in the motion video section which introduced advertising, but in the sequence on being accepted or rejected for a bank loan, the computation of expenditures, and the monthly profit statement.

Another condition which led to the success of this project was the collaborative approach. Although both designer and producer took primary responsibility for the tasks that traditionally fall into their areas, there was a continual exchange of ideas between the areas. That is, the producer suggested instructional techniques, and the designer offered ideas for production.

Summary

The experiment was judged a success. The agency learned that it could apply its expertise with linear video to videodisc technology. The *Business Disc* continues to be distributed nationally and internationally for use in many settings. □

Part Four

Research and Evaluation of Interactive Video

Interactive Video and the Control of Learning

Diana M. Laurillard

Introduction

The combination of computer-assisted learning (CAL) with video provides an inviting new educational medium that educational technologists are eager to exploit. Interactive video has obvious potential because it brings together a good expository medium with a good interactive medium. The coupling is not straightforward from the educational point of view, however, and poses a number of difficult questions for the designer to resolve. Some of these are amenable to research, and this article describes a feasibility study designed to investigate the new medium and to provide some information about the kind of learning experience it offers to students.

The Research Issues

The Balance of Control over Content Sequence and Learning Strategy. Many existing forms of interactive video (IV) have been derived from the addition of video to CAL and thus came from the programmed learning tradition. The effect of this kind of design is to lock the student into a particular branching sequence determined by the correctness of his or her input. It is common to find IV packages that control both the content sequence (i.e., the level of detail or generality, the order in which topics are presented) and the learning strategy (i.e., the alternation between presentation of information and practice, remedial loops, practice questions, the timing of testing). A highly controlled package of this type may be seen as desirable, because it is assumed that the teacher is capable of designing a learning task that will be more efficient than one in which control is left to the student. But there are possible disadvantages in using program control over both sequence and strategy.

Student control of the content sequence is an important issue for the design of interactive video, because the videodisc provides vast information resources of an entirely different kind from those already provided by the computer. As the amount of information increases, so also does the number of ways in which it can be put together, and ideally, we should make full use of this potential. It is possible to achieve this with student control, using a flexible interfacing facility that allows students free access to all the material, at any time, in any sequence they wish. Program control over the route through that material necessarily defines a subset of the possible uses of it, and greatly reduces its potential value.

There is the additional problem that in free learning situations (such as those described by Mager, 1961, and Pask and Scott, 1972), students exhibit a variety of alternative routes through the same material, very few of which ever correspond to the kind of logical route a teacher is likely to select. The students' route is likely to be one that is meaningful for them, and that means that any alternative is likely to be less meaningful. Program control over the content sequence may therefore be less educationally efficient than student control.

Program control over learning strategy presents a similar dilemma for the designer. There has been a continuing controversy in the literature about the efficacy of student versus program control over learning strategy, from Fry (1972) to Hartley (1981), and evidence has been produced on both sides. The more recent, and the more convincing of these is the later study, which suggests that student control over strategy is the more educationally effective. If this is so, then the design effort should be toward developing suitable forms of student control of strategy, rather than allowing this to be the exclusive responsibility of the program.

The balance of control over both sequence and strategy is an issue that must be addressed by interactive video designers, and needs further investigation to find out how to determine the optimal balance.

The Alternation Between Modes of Learning. The second issue for consideration by designers concerns the feasibility of combining an expository medium with an interactive one. From the students' point of view, interactive video requires them to move constantly between an active mode within the computer-driven part, and a receptive mode as they watch the video. Video is not necessarily a passive medium; there is evidence that students are extremely active as they watch an educational television program (Laurillard, 1982a). But it does force the students into a *receptive* mode of learning where they are invited to follow the presentation without themselves initiating any thinking processes. When the video stops and switches into the interactive part, the program will,

Diana M. Laurillard is with the Institute of Educational Technology, The Open University, Milton Keynes, England.

if it is any good, invite the students to become active, to initiate thought processes such as problem-solving, calculation, planning, executing a procedure, etc. This alternation between receptive and active modes of learning is precisely what makes IV an attractive medium, but it is not obvious how best to exploit it. It is a feature of good textbooks, or text-based teaching materials that they interrupt the text frequently with exercises or activities that allow the students to rehearse or actively process the information presented. But there is little research on the efficacy of this practice, and what there is suggests that students may often ignore the activities if they do not fit their personal strategy (Elton and Hodgson, 1976), or, if they do them, they become very instrumental in their processing of the text, attending only to those aspects that immediately address the exercise questions (Marton, 1974). Therefore, we can expect little help from traditional educational technology on this issue.

Moreover, there is an additional complication. Unlike text-based teaching materials, interactive video tends to give students less immediate control over the alternation between active and receptive modes. Students learning from text may choose among several different strategies for dealing with in-text questions: They may ignore them altogether, return to exercises at the end, begin by looking at what the exercises ask for, read beyond each exercise and then return to it, etc. But this degree of choice tends not to be available with the computer-controlled media—if a computer asks you a question, nothing will happen until you answer it. Again, this is seen as one of the virtues of this type of medium, but it is only a virtue if it is not destructively intrusive into the student's own approach to the task. If we were confident that our own pre-ordained structuring of exposition followed by practice was optimal for each student, there would be no problem. But the evidence from text-based teaching materials cited above suggests that when they have the opportunity, students tend to override these careful structures. Whether students or teachers know best is a moot point. But this issue of the control over the balance between receptive and active modes of learning is important for interactive video. We need to know more about how students experience the two and the transition between them.

Definition of the Research Questions. The three aspects of control defined above can be formulated as three basic research questions: What should be the balance of student control versus program control with respect to (a) the sequence of presentation of content, i.e., the strategy for receptive mode, (b) the choice of how many practice exercises to do, at what level of difficulty, i.e., the strategy for active mode, and (c) the strategy for alternating between the two modes of receptive and active learning? The research study described below was designed to investigate these questions.

The Interactive Video Teaching Materials

For the purposes of a research study, it was necessary to put together an interactive video program that was (a) not too costly, and (b) a viable teaching package. The first criterion meant that it was important to use existing video, rather than specially shot material, and as there is far more high-quality educational video available on cassette than on disc, the medium chosen was computer-controlled videocassette.[1] The second criterion required that the material should be based on an existing teaching package that had been carefully designed and evaluated, and was known to be effective. It was also important to choose a subject area that lent itself to the medium, i.e., for which a full understanding required both exposure to computer visuals, and rehearsal of procedures that are definable algorithmically. All these criteria were satisfied by a module called "Signals in Communication," part of the first-year course in technology at the Open University, T101: "Living with Technology." The module consisted of a 25-minute television program plus a 20-page text with diagrams and exercises, plus a one-hour audiocassette with diagrams and exercises.[2] A comprehensive evaluation study of all these teaching materials showed that the television program and audiocassette exercises especially had been very successful (Brown, 1981). The design of the interactive video package, therefore, made use of the television program to provide the video component, and the exercises to provide models for the CAL component.

The TV program had a sequential structure, and was didactic in form, which meant that it was possible to restructure it as a series of self-contained instructional sequences.

For each concept covered by the program, a CAL program was written to provide practice exercises, with tutorial guidance where necessary, and tests of the student's understanding of that concept. The form of these was based on the form of the exercises in the original text, with the embellishments that interactive computer graphics allowed. The teaching provided by the CAL programs did not go beyond that already contained in the original text.

With the design of the new material based so closely on already proven material, it was hoped that the criterion of viability would be met.

The critical design question, however, was: How to combine the two media? In order to investigate the three research questions already defined, it was necessary to choose some particular balance of control between student and program. For research purposes, the most revealing information is provided by the design format that gives maximum control to the students. By allowing them maximum control and seeing where this breaks down and why, we can determine what degree of control students can cope with, and what they need to be provided by the program.

The package was thus designed to give the students control in the three ways defined above. Figure 1 shows how the package was structured. The *control of mode* was given by allowing the students to choose exposition (video index) or practice (CAL index) at the start at A. If they chose CAL, at the end of the selected program they were again allowed to choose between modes at A. If they chose video, the selected sequence was presented. Here there was a difference between the two modes. In each CAL program, there is a natural choice point at the end of each exercise, when the student is allowed to choose either another exercise, or to abort the program and do something else (at C). No such natural choice point occurs in the video unless it is programmed to stop at the end of the exposition of a particular concept. To give the students maximum choice over *when* they were to change mode, the video was programmed to continue playing until interrupted at E. This meant that students could watch it continuously if they wished. When they did interrupt, they were given the choice of continuing from where they stopped, or moving a few seconds forwards or backwards, or choosing another sequence, or choosing the CAL mode (at F). In the second half of the student trials, the interrupt facility was removed and the video was programmed to stop at the end of each scene, giving the students the same choice as at F, and making the format of the video similar to that of the CAL. The comparative results of these two designs are reported below.

The *control of sequence* of presentation of the various concepts contained in the video was done via the video index, at D. The *control of learning strategy* was given in two ways: in what to practice, by selecting from the CAL index at B, and in how much to practice, by deciding on how many exercises to do, at C.

With the teaching materials so designed to give students the maximum control over the sequence of content, the alternation between exposition and practice, and the amount and timing of practice and testing, it was possible to set up realistic field trials for investigating the issue of the balance of control in interactive video.

Field Trials

Realistic field trials are logistically difficult at a distance learning university. It is only possible to bring sufficient numbers of students together for a trial at summer school. To run a realistic test of the material, however, it should be given to them in April at the appropriate point in the course. By using a group of summer school students, it was inevitable that for many of them this was a review exercise of material covered five months previously. It is possible to omit sections of a course, however, and for some students this was, in fact, an initial learning experience. The field trials at summer school, therefore, provided a valid, comparative study of the medium for both initial learning and review purposes, but did not provide extensive data on the former.

The field trials were carried out with 22 students at the summer school for the first-year technology course. The equipment was set up in an annex to a microprocessor laboratory, where the students were carrying out a one-day self-paced practical exercise. This set-up made it possible for students to be recruited at intervals throughout the day, to try out the interactive video teaching package. Each trial session lasted 30-60 minutes. Students were not expected to go through all the material. The aim was to observe learning behavior in terms of the strategies and responses of several different students confronted with this learning medium.

Two main trial runs were conducted with eight students (including three pairs) in the first, and 14 students (including three pairs) in the second. The main difference (apart from minor program debugging) between the two trial runs was the change to the interrupt facility described above.

Each trial began with a short computer-based pretest, covering the main concepts in the unit. Students were told only whether their answers were right or wrong and were given a final score at the end. The students were then started on the teaching material, and allowed to work through it as they wished. When they had finished, they were asked, "What did you think of this method as a way of learning this kind of material?" All student inputs and comments were recorded by an observer for later analysis, and results are outlined in the next section.

Results of the Field Trials

The results described below are based on computer-monitored data from the pretest, and observer-monitored data on students' choice of mode, sequence, and strategy as exhibited by their inputs,

Figure 1

Structure of Interactive Video Package, Mark 1

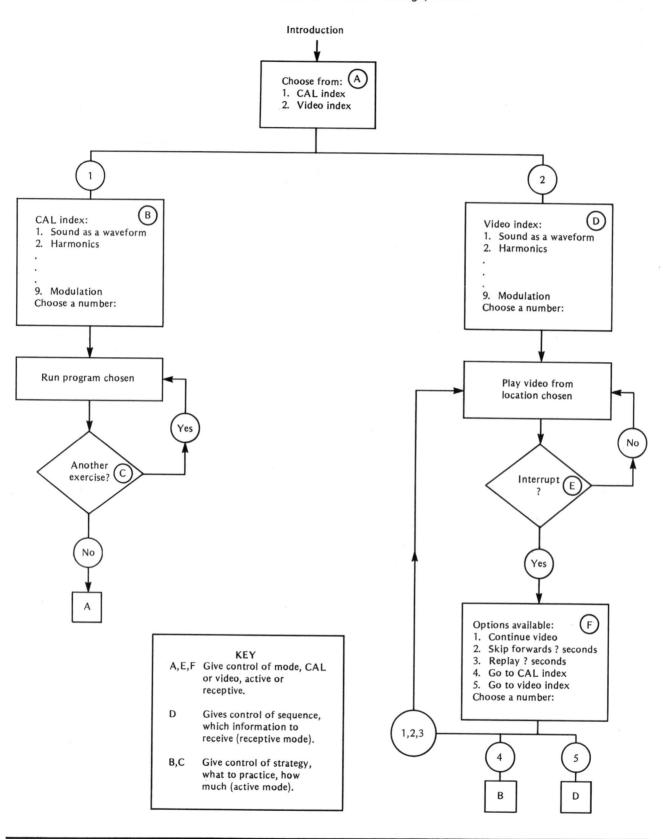

the time taken, and the comments made at the end.

Pretest Results. Many of the students in the field trial had already been exposed to the subject matter earlier in the year, and had seen the TV program when broadcast. This was reflected in the results of the pretest. The scoring was generally poor, however: Out of a total of nine questions, the average score was 4.7, ranging between zero and nine. Twelve of the students had either not covered the material before, or had felt confused by it. Their scores were all below 4.5, and they tended to treat the session as a learning session rather than as review. The remaining ten students scored five or more, and for them, much of the material was review.

Tolerance of Search Time. One problem in using videocassette equipment is that on requesting a video sequence, search times can take many seconds, whereas the maximum search time on a videodisc is approximately four seconds. It was useful to put up a frame describing what was happening as the VCR machine searched for the sequence requested. This reassured the students that they were not expected to do anything but wait. They were surprisingly tolerant of search times up to 20 seconds (the longest experienced), and none complained of this in the interviews. Whether this degree of tolerance would survive repeated use of the medium needs to be tested.

Control of Mode. One consequence of the decision to give students maximum control over their learning was that they should not only be allowed to choose what to do, but also when they should make this choice, i.e., they should be allowed to interrupt at any stage and be given the full choice of all options. When this is possible, the learning medium most nearly approaches the considerable flexibility that a book offers. However, it is difficult to achieve this with CAL programs. In BASIC, it requires an "ON ESCAPE GO TO . . ." statement at the beginning of each program, and students must be trained first to press ESCAPE each time they wish to interrupt. On some microcomputers, this can be a rather hazardous solution, and was avoidable here, as students were given the choice to escape from the program at the end of every exercise, and exercises were quite short.

The interrupt facility was more easily achievable on the video, however, and this allowed us to investigate two conditions: student-initiated interrupt, and program-initiated interrupt. In the first condition, while a video scene is playing, the computer program executes a loop to check whether a keyboard key has been pressed, and if so jumps out of the loop, stops the video, and puts up an option list (see E,F, in Figure 1). If no key has been pressed, the video continues playing until the end of the program, and then stops. The students are told at the beginning that they can interrupt the video by pressing a key, and this will allow them to move around the video (e.g., replay the last 30 seconds), or to do something else.

The second condition works in exactly the same way except that the video stops automatically at the end of each of the nine sequences, and puts up the same option list.

The comparative measure for these two groups was taken as the average length of time spent on one mode (receptive or active: video or CAL) before changing to another. The first condition, measured for four sets of students, gave an average of 12 minutes before switching modes; the second condition, measured for seven sets of students, resulted in an average of eight minutes per mode. This difference was accounted for entirely by the fact that in the first group, the students tended to continue watching the video, after it had moved to a different topic than the one requested. The students tended, therefore, to accept what was being given them, in receptive mode, rather than take the initiative to break away, even though the video had continued to a different topic. This is, of course, inevitable in a medium that is expressly designed to hold the attention and to smooth the transition from one topic to another. A similar finding occurs in text material, where students treat the text in the same linear fashion in which it is written, and tend not to break out of this to follow their own objectives. It is thus not surprising that students tend to relinquish control in the receptive mode. The question is: Would it be preferable to force them to make a choice at regular intervals? Is the second condition educationally preferable?

One way to consider this question is to compare the behavior of the two groups to see whether they do in fact take the opportunity to change mode when it is offered. For the first condition, the percentage of occasions when students interrupted the video spontaneously at the end of a topic was 30 percent of the total number of possible occasions. For the second condition, when the video stopped at the end of a topic and offered a change of mode, this figure rose to 80 percent: on only 20 percent of these occasions did students choose to continue with the video.

The conclusions drawn from these data must be tentative because the numbers of students in each group are small (eight in the first and 14 in the second). However, the considerable difference in the behavior of the two groups supports the suggestion that students watching the video have indeed relinquished control over their strategy, and will only pursue it when offered the opportunity to consider what it should be. This means that while we should offer students the opportunity to interrupt at any time, where possible, we should also force them to make conscious choices at regular intervals, i.e., the second condition is likely to be educationally preferable.

This does not mean that we have to force students to alternate between modes—this decision should be part of their own learning strategy and will be discussed later in this article. The conclusion here is that they should be forced to make the decision.

Control of Sequence. This aspect of the study investigated how far students like to be advised on the sequence of content to be studied, and how far they like to make their own choice. For the first group of students, they were given no advice on the sequence of content: At each choice point, they were presented with the full video and CAL indexes from which to choose. The students commented that they liked this freedom, but also would have liked advice.

The second group of students was given advice on an appropriate sequence at each choice point. At the end of a video sequence, they were told which was the associated CAL program, and at the end of each CAL program, it was suggested which video sequence they should look at next. Of the total number of choice points encountered by this group, the students followed the expected content sequence on 48 percent of the occasions. The rest of the time, they made their own decisions about the appropriate sequence. Strategies varied: Some began with content they found easy; some went straight to the content they had difficulty with on the test; some worked through the material sequentially.

From these data, we may conclude that it is important to allow students to make their own choice of sequencing of content, as at least half the time this differs from the expected sequence. Students also like to have advice on what to do next, however, and do make use of it when offered.

A second aspect of control of sequence concerns the facility to move around the video. While a sequence is playing, it is possible to interrupt at any point, and move forwards or backwards a specified number of seconds. Only one student made use of this facility. Again, it requires that students take initiative during receptive mode, which as we have seen, they tend not to do, and this may be why it was used so little. Whether this is an important facility will have to be judged on the basis of more extensive trials.

Control of Strategy. The students' control of their learning strategy can be investigated with respect to two different aspects of their behavior: how far they follow the expected pattern of watching the video sequence, then doing the associated CAL program, then the next video sequence, etc., i.e., how far they alternate between the two modes of learning; and how they make use of the option to repeat CAL exercises.

The former assumes that students will wish to receive some input about a topic before trying the exercises on it, and having completed those exercises will wish to move on to the next topic. Advice is given in accordance with this. However, of the total number of occasions when students encountered this advice, they accepted it only 28 percent of the time. For the remainder, they either chose entirely different content, as discussed previously, or they used the converse strategy of trying the exercises first, and moving to the video only if they felt they had done poorly. It seemed, therefore, that the students did want to alternate between receptive and active modes, but not necessarily in the expected sequence.

Interestingly, those students who were using the session for initial learning (the 12 who scored four or less on the pretest) used the video twice as much as those using it for review: This is largely because the former group inevitably needed explanatory help from the video more often. The degree of alternation between receptive and active modes can therefore be expected to relate to the student's purpose in using the medium.

The second aspect of control of strategy gave students the option to decide how many exercises they should do on each CAL program. Most of them commented that they liked having this choice. Indeed, this seemed to be important, because students who had no difficulty in answering the first exercise would usually abandon that program immediately; others who were less certain would repeat three or four exercises to make sure they understood it; those who had real difficulty needed as many as 12 exercises before they felt confident with it.

For both these aspects of learning strategy, therefore, student behavior was unpredictable, and varied considerably among students.

Conclusions

The study was designed to investigate how students would operate within a complex learning medium that gave them maximum control over the learning process. We can conclude, albeit tentatively given the small numbers involved, that students can indeed make use of this maximum control. Learning behavior varies sufficiently among students, and differs

Figure 2

Modified Structure of Interactive Video Package, Mark 2

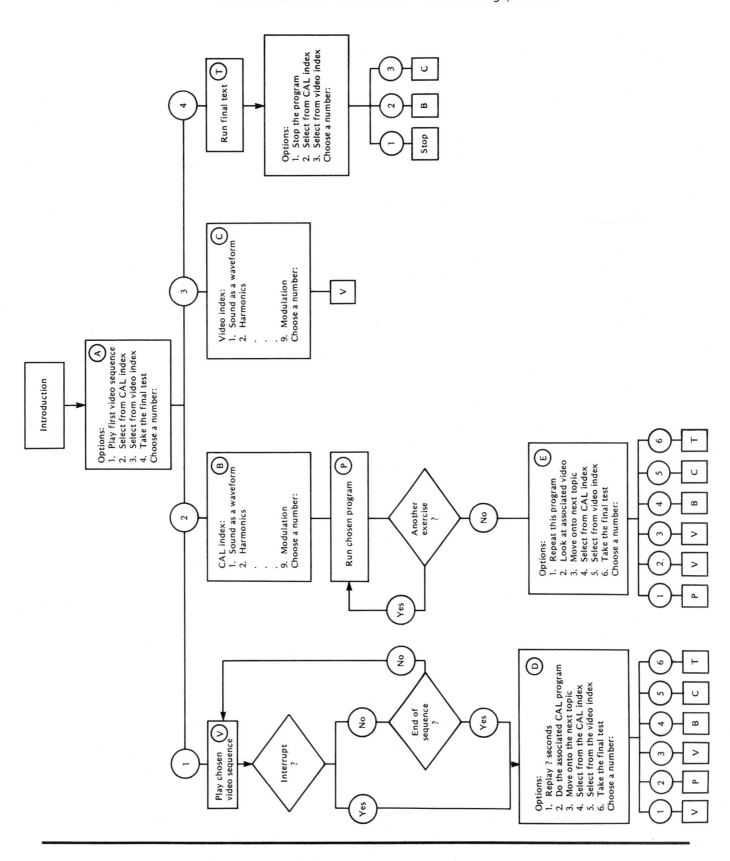

sufficiently from any expected learning strategy, that this degree of control becomes necessary if individual students' preferences are to be considered. Students also like to have advice on some suggested sequence and strategy, however, and do make use of it when offered. Finally, it would seem that students are unable to make full use of the control of *mode* of learning. They do alternate between the two modes, but they have to be forced to make this choice at regular intervals, within the receptive mode. It would be interesting to investigate whether they could make use of the interrupt facility if it were offered within active mode, i.e., if they could interrupt the CAL programs at any point.

The study has not attempted to relate student behavior in interactive video to differences in learning outcomes. This is because the learning task here is so complex that it needs an extremely sensitive evaluation instrument to be able to determine reliably significant differences in scores, and to correlate these with their source in the chosen learning strategy. This would require further research and goes beyond the immediate goal of determining the degree of control that students can make use of in this medium.

Discussion

The students in this study liked interactive video as a learning medium. In the interview, most commented that they enjoyed learning this way, and liked the degree of choice available to them. The feasibility of the medium is by no means obvious—students are confronted with an alarming array of equipment, they must master use of the keyboard, simple though it is, and they have to adapt to a medium that is quite unlike any they have met before. It withstood this particular test, however, and the students' experience was that it was an enjoyable and efficient way of learning.

The aim of the study was to examine students' behavior with the medium, and to consider the implications of this for the way we should set about the design of interactive video. One important aspect of design is the balance of control between the program and the student. Can the student handle total control? How far should the program take over?

The overwhelming conclusion from this study is that students can make full use of most aspects of control, and moreover make use of it in such a variety of ways that it becomes clear that program control must seriously constrain the individual preferences of students. To justify the use of program control, the designers must demonstrate that they know best what the student needs at each stage, i.e., that program control gives improved learning outcomes. Given the continual failure of educational research to ever find an evaluation instrument sensitive enough to produce significant differences of this kind, it must be preferable to give students the benefit of the doubt. We should acknowledge that the unpredictability and variation in their learning behavior could be derived from perfectly legitimate and effective learning strategies, and that these should be considered in the design.

The structure of the interactive video package as a whole has been refined as outlined in Figure 2. The main differences, introduced as a result of the study, are as follows:

1. Advice on what to do next is offered at A1, D2, D3, E2, and E3.
2. A final test has been added to allow students to gauge their knowledge.
3. The option to take the final test appears at every choice point A, D, and E.
4. The video automatically stops playing at the end of each sequence, rather than waiting for the student to interrupt.

The last feature is the only curtailment of freedom introduced as a result of the field trials. The conclusion from the study, as suggested previously, is that while in receptive mode, e.g., watching a video presentation, students tend not to initiate decisions about their learning strategy. It may be a mistake, therefore, to attempt to emulate the fullest flexibility that a book offers, by allowing the students to work continuously in receptive mode if they wish. The ability of a computer-controlled medium to stop and force the students to consider their strategy may be a virtue, rather than an imposition. However, this is as far as the imposition should go. If interactive video were to become as highly controlling and directive as most CAL has been, it would be a gross misuse of the medium, partly because it diminishes the potential variations in the use of stored data, and partly because it undermines the students' own responsibility for their learning. Giving the balance of control to the students has not only the benefit of being more democratic, it is probably also more educationally effective. □

Notes

1. The configuration for this equipment is described in more detail in Laurillard (1982b).
2. The module was designed by Professor John Sparkes, who wrote the text, and made the audiocassette package. The television program was produced by Phil Ashby and presented by Dr. David Crecraft.

References

Brown, S. Signals Tributary, Part V of Evaluation Report on T101: Living with Technology, IET, Open University, 1981.

Elton, L., and Hodgson, V. Individualised Learning and Study Advice. In J. Clarke and J. Leedham (Eds.), *Aspects of Educational Technology X, Individualised Learning.* London: Kogan Page, 1976.

Fry, J.P. Interactive Relationship Between Inquisitiveness and Student Control of Instruction. *Journal of Educational Psychology*, 1972, *63*(5), 459-465.

Hartley, J.R. Learner Initiatives in Computer Assisted Learning. In U. Howe (Ed.), *Microcomputers in Second-ary Education.* London: Kogan Page, 1981.

Laurillard, D.M. D102 Audio-Visual Media Evaluation, IET Papers on Broadcasting No. 212, Open University, 1982a.

Laurillard, D.M. The Potential of Interactive Video. *Journal of Educational Television*, 1982b, *8*(3), 173-180.

Mager, R.F. On the Sequencing of Instructional Content. *Psychological Reports*, 1961, *9*, 405-413.

Marton, F. Some Effects of Content-Neutral Instructions on Non-Verbatim Learning in a Natural Setting. *Scandinavian Journal of Educational Research*, 1974, *18*, 199-208.

Pask, G., and Scott, B.C.E. Learning Strategies and Individual Competence. *International Journal of Man-Machine Studies*, 1972, *4*, 217-253.

An Analysis of Evaluations of Interactive Video

James Bosco

Much has been written about interactive video during its ten-year history. The past five years have marked an especially abundant period of articles and reports about interactive video. Most of the literature falls into three groups. One group contains explanations about the nature and use of interactive video. A second group contains anecdotal descriptions of particular projects. The third group contains evaluations of interactive video. This is an important collection of reports, since it should provide objective information about how well interactive video works and how to make it work better. Such information can be very helpful for persons who are considering or developing interactive video projects.

The purpose of this article is to analyze the third group, the evaluation literature on interactive video. The reports included in this analysis met the following criteria: (1) All reports included pertain to the use of interactive video in instructional applications. Point-of-sales and archival applications were not included. (2) All were data based. Reports which contained only anecdotal or subjective information were eliminated. (3) All reports were in the public domain. Proprietary reports, even those for which secondary source information has been published, were not included.

The analysis of the evaluations is contained in four sections. The first section consists of a description of the nature of the evaluations of interactive video. This is followed by a review of information on the effectiveness of interactive video. In the third section, there is a discussion of what these evaluations offer for the improvement of interactive video design and use. The article concludes with a discussion of the implication of these evaluations for the conduct of future evaluations.

James Bosco is Director, Merze Tate Center for Research and Information Processing, College of Education, Western Michigan University, Kalamazoo, Michigan.

Nature of the Evaluation

One of the often heard comments about the evaluation of interactive video is that such evaluations are scarce. *Such is not the case.* This analysis was based on 28* reports, and the writer is aware of a number of other evaluation reports which were not obtained in time to be included here. Relative to other instructional techniques, and especially techniques which have been in general use for only about five years, the number of evaluation reports which have been produced is considerable.

The perception of a scarcity of evaluation reports may be a consequence of the difficulty in getting access to them. Only eight of the evaluations were published in journals; five were available through ERIC; five were presented at conferences; and the remainder were not available outside the agencies conducting the research. Some of these unpublished documents were referred to in other publications. Most were obtained as a result of leads which emerged in conversations with people around the country. Many phone calls were often necessary to find the agency or person who could make a report available. Even the evaluations in journals may be somewhat less conspicuous because they are published in diverse journals with quite different readerships.

Within what types of organizations were these interactive video evaluations produced? Table 1 shows the distribution of types of organizations wherein the interactive video took place. The predominant types of organizations wherein these evaluations were done were the military, higher education, and K-12 education. Social service agencies and industry were the least represented in the collection of evaluations in this report. The small number of studies from the private sector might be surprising, since the private sector, along with the military, has been the largest producer of interactive video. The low number of evaluations of interactive video from industry may reflect the proprietary nature of such evaluations; or, the low number may simply mean that there are fewer systematic evaluations done in the private sector. It is difficult to assess how many substantial and credible evaluations are being withheld from the public domain because of proprietary concern, in contrast to the number of less credible evaluations which may have more utility if they

*Subsequent discrepancies (28 or 29) result from the Brown report, which consisted of two evaluations combined into one report.

Table 1

Table 1

Distribution of Organizations Which Are
Represented in the Reports in This Analysis

Organization	Number of Evaluations
Military	10
Higher Education	8
Elementary Education	4
Junior High	2
High School	2
Industry	2
Social Services	1

(Note: One report has two sections and each section pertained to a different organization.)

are referred to but not made available for careful examination.

The evaluations which were reviewed for this analysis were both formative and summative. In a number of cases, both elements were involved in the evaluation. Although both disc and tape interactive video systems were included, most of the evaluations which were obtained used discs. With regard to the sample size, the studies ranged from 5 to 700 subjects. The duration of the interactive video instruction usually was a few days.

Information on Effectiveness

What do these evaluation reports reveal regarding the effectiveness of interactive video? Is interactive video an effective instructional technique? Table 2 contains a general summary of the basic aspects of the evaluations. This table presents the findings from each study. Since only outcome variables (i.e., effectiveness variables rather than process variables) are contained in this table, there are a number of N/A (non-applicable) entries. The other tables in this paper are derived from Table 2.

Table 3 lists the outcome variables used in these studies. As this table indicates, achievement, user attitude, training time, and performance were the typical variables used to measure outcomes. Achievement included paper and pencil tests and tests built into the interactive video presentation,

usually in a "real life" application. Performance involved the demonstration of the skill which was being taught by the interactive video presentation. The single use outcome variables in Table 3 generally were found in the Hull (1984) study.

The typical evaluation involved an experimental design with a treatment and a control group. The control group usually received the instruction which the disc or tape was intended to replace. The statistical procedure used most frequently was an analysis of variance or t test. In a couple of cases, regression analysis was used.

Sixteen of the 29 evaluation authors drew a "bottom-line" assessment or general conclusion regarding the effectiveness of interactive video. In 13 of the 16 instances, the general conclusion was that interactive video was effective. In three evaluations, the author concluded that interactive video was not effective and, in the other 12 evaluations, the author chose not to make such an assessment. The tone of the majority of these reports suggested that the general conclusion, had one been presented, would have favored interactive video.

Another way to see what these studies as a group tell us about the effectiveness of interactive video is to examine the "reported benefit" of the outcome variables. Reported benefit means that the evaluator claimed interactive video produced more favorable outcomes (e.g., higher achievement, more positive attitude, faster learning time, etc.) than the comparison. There were 39 instances involving the use of statistical tests when the evaluation reached a conclusion on benefits; i.e., interactive video either *did* or *did not* result in a more favorable outcome. In 24 of the 39 instances (61%), there was a positive finding for interactive video. There were 23 instances which did not involve the use of a statistical test when the evaluation reached a conclusion. In 22 of the 23 instances (96%), there was a positive finding for interactive video. Thus, when a statistical test was used, both positive and negative findings are reported, but when a statistical test was not used, negative findings were reported in only one instance. The use of statistical tests tended to result in more conservative conclusions about the benefit of interactive video.

Table 4 breaks out the findings for the four most common dependent variables: training time, achievement, performance, and user attitude. When the reported benefits involving the use of a statistical test were examined, benefits were most prevalent on user attitude and training time variables. They were less prevalent on the achievement variable. The achievement variable was only one case away from a 50-50 split (i.e., 7 to 5). No pattern emerged on the performance variables.

While the issue of differences in mean training

Table 2

Summary of Evaluations of Interactive Video

Principal Investigator and Publication Date	Instruction for:	Content	Video Storage	Type of Evaluation[1]	n	IV Compared With:	Outcome Variables	Reported Benefit	Stat. Tests	Where Pub.[2]	Amount of Info. on IV Design[3]
Andriessen (1980)	Higher Education	"Work of the Heart"	Disc	F S	12	No comparison	a-User attitude	a-Yes	a-No	J	L
Brown (1980)	Elementary Education	Tumbling	Disc	F	4 Classes	N/A	N/A	N/A	N/A	U	M
	Higher Education	Spanish	Disc	F	16	N/A	N/A	N/A	N/A	U	M
Holmgren (1980)	Military	Equipment use	Disc	S	298	Standard instruction	a-Performance test b-Training time c-User attitude	a-Mixed b-No c-Mixed	a-Yes b-Yes c-Yes	J	L
Yeany (1980)	Higher Education	Biology	Tape	S	99	Lecture and lab without IV	a-Achievement	a-Yes	a-Yes	C,E	L
Allen (1981)	Junior High	Tumbling, Spanish, Metrics	Disc	F	5 Teachers	N/A	N/A	N/A	N/A	U	L
Daynes (1981)	Junior High	Tumbling	Disc	F	5 Instruct.	N/A	N/A	N/A	N/A	J	L
Young (1981)	Military	Equipment Service	Disc	S	51	Practice using hands-on equipment	a-Objective test b-User-estimated confidence c-Performance test d-Retention e-Training time	a-Yes b-Yes c-Yes d-Yes e-Yes	a-Yes b-Yes c-Yes d-Yes e-Yes	U	E

1 - Type of Evaluation
F - Formative
S - Summative

2 - Where Published
J - Journal
U - Unpublished
E - ERIC
C - Conference

3 - Info. on IV Design
L - Little or none
M - Moderate
E - Extensive

(Continued)

Table 2 (Continued)

Principal Investigator and Publication Date	Instruction for:	Content	Video Storage	Type of Evaluation [1]	n	IV Compared With:	Outcome Variables	Reported Benefit	Stat. Tests	Where Pub. [2]	Amount of Info. on IV Design [3]
Gibbons (1982 b)	Military	Equipment Service	Disc	F	N/A	N/A	N/A	N/A	N/A	U	M
Gibbons (1982 a)	Military	Equipment use	Disc	F,S	Study 1 7 Study 2 2 Instruct. Study 3 46	Study 1 No comparison Study 2 No comparison Study 3 Normal classroom instruction	Study 1 N/A Study 2 N/A Study 3 a-Post test b-User attitudes c-Solution time	Study 1 N/A Study 2 N/A Study 3 a-Yes b-Mixed c-Yes	Study 1 N/A Study 2 N/A Study 3 a-No b-No c-No	U	L
Hofmeister (1982)	Elementary	Color matching	Disc	F	11	N/A	N/A	N/A	N/A	C,E	L
King (1982)	Military	Equipment service	Disc	S	235	Self-paced instruction with lab	a-Progression index (training time) b-Retention c-Number of users repeating module d-User acceptance e-User attitude f-Instructor acceptance	a-Yes b-No c-No d-Yes e-Yes f-Yes	a-Yes b-Yes c-Yes d-Yes e-Yes f-No	U	L
Thorkildsen (1982)	Elementary Education, Adults	Special education	Disc	F,S	Study 1 6 Study 2 ? Study 3 ? Study 4 7 Study 5 32	Study 1 No comparison Study 2 No comparison Study 3 Tutoring Study 4 No comparison Study 5 Alternative IV structure	Study 1 N/A Study 2 N/A Study 3 a-Achievement Study 4 N/A Study 5 a-Paper and pencil test b-IV test c-Training time	Study 1 N/A Study 2 N/A Study 3 a-No Study 4 N/A Study 5 a-No b-No c-No	Study 1 N/A Study 2 N/A Study 3 a-Yes Study 4 N/A Study 5 a-Yes b-Yes c-Yes	E	E

(Continued)

Table 2 (Continued)

Principal Investigator and Publication Date	Instruction for:	Content	Video Storage	Type of Evaluation[1]	n	IV Compared With:	Outcome Variables	Reported Benefit	Stat. Tests	Where Pub.[2]	Amount of Info. on IV Design[3]
Thorkildsen (1982) (continued)					Study 6 N/A	Study 6 Cost of teacher aid	Study 6 a-Cost	Study 6 a-Mixed	Study 6 a-No		
Wilkinson (1982)	Military	Equipment maintenance	Disc	F,S	59	Lab with real equipment	a-Cost b-Performance test c-Time on performance test d-User acceptance	a-Yes b-No c-No d-Mixed	a-No b-Yes c-Yes d-Yes	U	L
Gratz (1983)	Higher Education	Teacher evaluation	Tape	S	60	No instruction	a-Achievement	a-No	a-Yes	E	L
Henderson (1983)	High School	Mathematics	Tape	S	Study 1 81 Study 2 9	Study 1 Regular classroom instruction without IV Study 2 No comparison	Study 1 a-User attitude b-Achievement c-Self-assessment of "school learning" Study 2 a-Achievement	Study 1 a-Yes b-Yes c-No Study 2 a-Yes	Study 1 a-Yes b-Yes c-Yes Study 2 a-Yes	E	M
Kirchner (1983)	Elementary Education	"Heart and circulation"	Disc	F,S	35	No comparison	a-User Attitude	a-Yes	a-No	J	E
Wooldridge (1983)	Industry	Lift truck safety	Tape	S	16	No comparison	a-Post test	a-Yes	a-No	J	L

1 - Type of Evaluation
F - Formative
S - Summative

2 - Where Published
J - Journal
U - Unpublished
E - ERIC
C - Conference

3 - Info. on IV Design
L - Little or none
M - Moderate
E - Extensive

(Continued)

Table 2 (Continued)

Principal Investigator and Publication Date	Instruction for:	Content	Video Storage	Type of Evaluation [1]	n	IV Compared With:	Outcome Variables	Reported Benefit	Stat. Tests	Where Pub. [2]	Amount of Info. on IV Design [3]
Bunderson (1984)	Higher Education	Biology concepts	Disc	F,S	Study 1 35 Study 2 150	Study 1 Classroom lectures Study 2 Classroom lectures	Study 1 a-Achievement b-Training time Study 2 a-Achievement b-Learning productivity	Study 1 a-Mixed b-Yes Study 2 a-Yes b-Yes	Study 1 a-Yes b-Yes Study 2 a-Yes b-Yes	J	M
Davis (1984)	Higher Education	Biology Lab	Disc	F,S	688 Users 45 Instruct.	Traditional lab	a-Training time b-User attitudes c-Instructor attitudes	a-Yes b-Mixed c-Mixed	a-No b-No c-No	U	E
Glenn (1984)	High School	Economics	Disc	S	44	No comparison	a-Achievement b-Training time c-User attitudes d-Instructor attitudes	a-Yes b-Yes c-Yes d-?	a-No b-No c-No d-No	J	L
Hull (1984)	Military	Equipment use	Disc	S	72	Hands-on training with disc	a-9 written tests b-Practical exam c-Training time d-User's perceived confidence e-User's perceived speed of skill acquisition f-Instructor's perceived speed of skill acquisition g-User's perceived attention h-Instructor's perceived attention i-Perceived transfer of learning j-Instructor's perception of transfer of learning k-User acceptance l-Attitude toward practical experience	a-Mixed b-Yes c-Yes d-Yes e-Yes f-Yes g-No h-Yes i-Yes j-Yes k-No l-Yes	a-Yes b-Yes c-Yes d-Yes e-Yes f-No g-Yes h-No i-Yes j-No k-Yes l-Yes	U	L

(Continued)

Table 2 (Continued)

Principal Investigator and Publication Date	Instruction for:	Content	Video Storage	Type of Evaluation [1]	n	IV Compared With:	Outcome Variables	Reported Benefit	Stat. Tests	Where Pub. [2]	Amount of Info. on IV Design [3]
Hull (1984) (continued)							m-User attitude n-Program implementation	m-Yes n-Mixed	m-Yes n-Yes		
May (1984)	Industry	Equipment service	Disc	S	89	Self-paced instruction	a-Training time b-Performance test c-User attitude	a-Yes b-No c-Mixed	a-No b-No c-No	C	L
Parker (1984)	Higher Education	Human genetics and spina bifida	Disc	S	?	Traditional instruction	a-Cost	a-Yes	a-No	J	L
Smith (1984)	Social services	Training for personnel to administer Aid to Families with Dependent Children	Disc	S	Approx. 500	Traditional training	a-Achievement	a-Yes	a-No	C	L
Vernon (1984)	Military	Equipment operation	Disc	S	144	Classroom instruction with lab	a-Training time b-Performance c-Cost d-User acceptance e-Instructor attitudes	a-Yes b-Mixed c-Yes d-Yes e-Yes	a-No b-Yes c-No d-No e-No	U	L
Wager (1984)	Military	Keyboard skills	Disc	S	67	Linear video tapes and classroom exercises	a-Achievement b-Cost c-User acceptance d-Instructor acceptance	a-No b-Yes c-No d-No	a-Yes b-No c-Yes d-Yes	U	L
Huntley (1985)	Higher Education	Recognition of motor dysfunction in infants	Disc	F,S	130	Regular instructor without the disc	a-Achievement b-User attitude	a-Yes b-Mixed	a-Yes b-No	C	E
Ketner (n.d.)	Military	Equipment service	Disc	S	40	Self-paced video tape and lab	a-Training time b-User acceptance	a-Yes b-?	a-Yes b-No	U	L

Table 3

Outcome Variables Used in Evaluation of Effectiveness

Achievement —————————————————— 18
User attitude —————————————————— 16
Training time —————————————————— 11
Performance ——————————————————— 6
Instructor attitude ————————————————— 4
Cost ————————————————————————— 5
Retention ——————————————————————— 3
User-estimated confidence ——————————— 2
Number of users repeating module —————— 1
Time on performance test —————————————— 1
Self-assessment of school learning ————— 1
Learning productivity ————————————————— 1
User's perceived speed of skill acquisition —— 1
Instructor's perceived speed of skill acquisition —— 1
User's perceived attention ————————————— 1
Instructor's perceived attention ———————— 1
Perceived transfer of learning ———————————— 1
Instructor's perception of transfer of learning —— 1
Program implementation ————————————— 1

time was the primary focus of the analysis of training time, there is an accompanying issue of the differences in the standard deviation of training time factor. Interactive video tended to reduce mean training time but there was also a tendency for it to increase the standard deviation (Bunderson, Lipson, and Fisher, 1984; Smith, 1984). One rationale for interactive video is to accommodate differences among learners, and the large standard deviations for training time indicate that such is occurring; yet, as Smith points out, this might create a problem in some situations. In other situations, such as many industrial applications, the advantage of removing individuals from work for training offsets the reintegration problem.

In several studies an attempt was made to examine the impact of variables which might affect success on the interactive video training. Background variables such as amount of education, prior training, and age were explored in several studies (Holmgren *et al.*, 1970; Wilkinson, 1982; Wager, 1984; Wooldridge and Dargan, 1983). Cognitive or personality variables were considered in other studies (Yeany, 1980; Hull, 1984). These variables generally had no differential impact on outcome.

Results of Formative Evaluations

Even though nearly half of the evaluations involved a formative component, the studies were generally disappointing on this point. In many instances the formative evaluation was happenstance and quite casual. The typical approach to formative evaluation involved questionnaires or interviews with users.

The most extensive reported formative evaluation (in terms of sample size) was the Davis (1984) evaluation of the Annenberg ICPB Project to develop interactive videodiscs to simulate laboratory experiences in college-level physics, chemistry, and biology. Six hundred and eighty-eight users and 45 instructors were involved. Four major conclusions for the development of interactive video instruction programs emerged:

1. Build on features known to be effective from other instructional materials.
2. Make the structure of the disc obvious.
3. Provide the maximum amount of user control.
4. Use feedback messages to reinforce performance and provide diagnostic messages to correct errors (Davis, 1984).

While it is hard to dispute the reasonableness of these four points, it might be expected that a formative study as extensive as this one would produce a somewhat more trenchant analysis of the deficiencies in interactive video, suggesting specific ways to improve products.

The conclusions which have emerged from other formative evaluations tend to be at a similar level. It is difficult to see any appreciable difference between the formative evaluation information produced in 1980 and that which has been produced in 1985. This is unfortunate because the experiences of others should be helpful in the avoidance of mistakes by the person new to the development of interactive video. Perhaps one of the reasons for the limited utility of the formative aspects of the evaluation is that, in many of the cases, the interactive video project reported was the first, or one of the first, attempted. Under these circumstances it is not surprising that the information resulting from the formative evaluations provides only modest assistance, and that no growing level of sophistication is evident in the literature.

A few of the reports did provide useful formative information. Thorkildsen's (1982) evaluation provides an excellent example of how formative evaluation can be used to make improvements in the development of interactive video. In this project, formative evaluation was a carefully conceived aspect of the development of the instruc-

Table 4

Reported Benefits on Achievement, Training Time,
User Attitude, and Performance When Statistical Tests Were Used and Not Used

| | Benefit Reported | | No Benefit Reported | | Mixed Results | |
	Stat Test Used	Stat Test Not Used	Stat Test Used	Stat Test Not Used	Stat Test Used	Stat Test Not Used
Achievement	7	4	5	0	2	0
Training Time	5	4	2	0	0	0
User Attitude	5	4	2	0	2	3
Performance	1	0	1	1	2	1

tion and was actually used in the refinement of the design of the interactive video instruction. The Gibbons et al. (1982b) evaluation contains much helpful information about the production process and the instructional design process. This information should be particularly valuable to individuals producing their first disc. Bunderson et al. (1984) has a useful summary table of features individuals liked and disliked in the interactive video. Wager's (1984) report contains a useful discussion of the cost issue.

Implications of Previous Evaluations for the Design of Future Evaluations

The typical evaluation in the collection was focused on the question: Is interactive video more effective than traditional instruction? Effectiveness was assessed by achievement, performance, and attitude measures. When examined as a group, the studies in this report do not lead to a categorical answer. Interactive video did seem to result in positive findings on user attitude and training time, but the benefit of the faster training time needs to be considered along with the achievement outcome, which split at around 50 percent in favor of interactive video. In essence, sometimes interactive video appeared to be more effective than the comparison instruction, and sometimes it did no better than the comparison form of instruction.

Wide Variety

The major conclusion to be drawn from this is that it is not useful to think about interactive video as an approach but, rather, as a category designation for a wide array of approaches. The only shared aspect of various interactive video pro-

grams is that microprocessing control of video information provides nonlinear video information. By way of analogy, the question "Are books effective in providing instruction" does not lead to a categorical answer. Rather, the answer to this question depends upon the content of the book, the way it is being used, the objectives of instruction, etc. *The consequences of interactive video will depend on the design and use of specific programs.*

As the use of interactive video instruction increases, there ought to be a growing sophistication about how to make it work and a clearer sense about the ingredients in successful applications. One of the difficulties in attempting to "tease out" conclusions from the various evaluations of interactive video which might help explain success or failure of interactive video instruction is that few reports provide information on the specifics of the design. Among the evaluations contained in this report, 19 of the 29 provided little or no information on the nature of the interactive video program; 5 provided a moderate amount of information; and 5 provided extensive information. When information on the specifics of the instruction are lacking, it is impossible to look for clues which might explain differences in findings on effectiveness.

Another useful basis for examining differences would be a means of defining alternative types of interactive video instruction. The only generally agreed upon characterization of interactive video is the University of Nebraska taxonomy, which defines four levels of interactive video oriented to the hardware. If a good classification system for alternative interactive video instructional designs were available, it might be possible to discern the consequence of different approaches on outcome variables.

Criterion for Effectiveness?

Another issue concerns the question of the criterion for defining effectiveness. The most common approach to the criterion issue is to define effectiveness as the attainment of a higher score than the instruction which the interactive video may replace. Such a comparative approach has the potential to be misleading. If we do not know how effective the comparison instruction is, we may be misled in thinking we have an improvement. There is little value in marginal improvement, even if it is statistically significant, over a weak comparison.

Brody (1984) and Reeves (1984) provide thoughtful critiques of the research on interactive video and a variety of useful suggestions about improving the quality of research and evaluation of interactive video. Underlining their discussions is concern about how to collect data which advances understanding about the development of and use of interactive video instruction. While the experimental designs may provide a semblance of rigor in evaluations, they may be less useful than other approaches, especially at this stage in the development of interactive video. Reeves provides several suggestions about alternatives which persons involved in educational design would do well to consider.

One potentially useful, if underused, approach is structured observations. Three of the studies in the group made use of this approach (Davis, 1984; Bunderson *et al.*, 1984; Kirchner, 1983). This approach can be very useful in providing insight into the interaction of the learner with the instructional system. In Bunderson's study, for example, five alternative patterns of use of the system were observed. These patterns were the distinctive approaches of students using the interactive capability which was available. Understanding how individuals make use of the flexibility can be helpful in designing systems. While structured observation may not appear to have the rigor of an experimental design, careful observation can help to provide insight into the realities of use of interactive video. Such observations may be particularly useful when the designer follows the interactive video instruction into the field. The use of inexpensive test discs can enable designers to take advantage of any information obtained prior to pressing the master.

Little use was made of evaluation as an integral and unobtrusive component of the interactive video. One of the especially valuable elements of interactive video is that it is possible to track the user through the instructional sequences and to record salient aspects of the experience as it is occurring. This is an especially potent tool for design development which should be used more frequently.

Four Categories of Issues

The ultimate success of interactive video entails four categories of issues. First is the hardware issue. Clearly, the hardware which is available is good and continues to get better. Reliability has increased, cost has decreased, and increasing enhancements are being provided. The second issue is video production. Here, again, the needed capabilities are readily available. The third is the instructional design. There is a need for individuals who can approach interactive video with fresh ideas about instructional design, using the video and computer capabilities which interactive video provides. We have passed the point when the novelty of interactive video is sufficient to mask inferior instructional design. Evaluation is an important tool for those who seek to push the design of instruction on to new paths through the use of interactive video. The fourth issue is the integration of interactive video within the customs and procedures of an ongoing organization.

Very few of the studies involved awareness of the organizational problems in applying interactive video in real life situations. The question is, "How well does it do the job?" This question is only in part answerable with regard to the quality of the courseware. It also implies issues of integration of interactive video within the total organizational structure. There is need to look at the application in real world context which can involve other conditions present in the application site which may jeopardize or could possibly enhance the quality of the instruction.

Conclusion

Interactive video provides a new palette for persons involved in developing instructional programs. The challenge of interactive video is not to paint the same old pictures with new colors, but to paint new pictures. □

References

Allen, L. *Field Test Results of Nebraska-Produced Videodiscs.* The Nebraska Videodisc Project, Videodisc Design/Production Group, KUON-TV. Lincoln: University of Nebraska, 1981.

Andriessen, J.J., and Kroon, D.J. Individualized Learning by Videodisc. *Educational Technology*, 1980, *20*(3), 21-25.

Brody, P.J. *Research on and Research with Interactive Video.* Paper prepared for the annual meeting of the American Educational Research Association, New Orleans, 1984.

Brown, R.D., and Newman, D.L. *A Formative Field Test Evaluation of Tumbling and Spanish Videodiscs.* The Nebraska Videodisc Project, Videodisc Design/Production Group, KUON-TV. Lincoln: University of Nebraska, 1980.

Bunderson, C.V., Lipson, J.I., and Fisher, K.M. Instructional Effectiveness of an Intelligent Videodisc in Biology. *Machine-Mediated Learning,* 1984, *1*(2), 175-215.

Davis, B.G. *Science Lab Videodiscs: Evaluation Report and Critique of each Videodisc.* Unpublished manuscript, University of California, Berkeley, 1984.

Daynes, R., Brown, R.D., and Newman, D.L. Field Test Evaluation of Teaching with Videodiscs. *Educational and Industrial Television,* March 1981.

Gibbons, A.S. *Videodisc Training Delivery System Project: Final Technical Report.* Prepared for U.S. Army Training and Doctrine Command, Fort Monroe, VA, 1982a.

Gibbons, A.S., Olsen, J.B., and Cavagnol, R.M. *Hawk Training System Evaluation Report.* Unpublished manuscript, WICAT Systems, Inc., Orem, UT, 1982b.

Glenn, A.D., Kozen, N.A., and Pollak, R.A. Teaching Economics: Research Findings from a Microcomputer/Videodisc Project. *Educational Technology,* 1984, *24*(3), 30-32.

Gratz, E.W., and Reeve, R.H. *Individualizing Instruction Through Interactive Video.* Washington, D.C.: Pan American University, 1983. ERIC Document Reproduction Service No. ED 234 750.

Henderson, R.W. *et al. Theory-Based Interactive Mathematics Instruction: Development and Validation of Computer-Video Modules.* Santa Cruz, University of California, 1983. ERIC Document Reproduction Service No. ED 237 327.

Hofmeister, A.M. *The Videodisc and Educational Research.* Washington, D.C.: Department of Education, 1982. ERIC Document Reproduction Service No. ED 236 218.

Holmgren, J.E., Dyer, F.N., Hilligoss, R.E., and Heller, F.H. The Effectiveness of Army Training Extension Course Lessons on Videodisc. *Journal of Educational Technology Systems,* 1980, *8*(3), 263-274.

Hull, G.L. *An Evaluation of the Student Interactive Training System at the U.S. Army Air Defense School. Final Report.* Prepared for U.S. Army Communicative Technology Office, Fort Eustis, VA, and U.S. Army Air Defense Artillery School, Fort Bliss, TX, 1984.

Huntley, J.S., Albanese, M.A., Blackman, J., and Lough, L.K. *Evaluation of a Computer Controlled Videodisc Program to Teach Pediatric Neuromotor Assessment.* Iowa City, IA: University of Iowa. Paper prepared for the American Educational Research Association, Chicago, 1985.

Ketner, W.D., and Carr, E.P. Unpublished manuscript. *A Training Effectiveness Analysis of Standard Training Techniques and the Videodisc Microprocessor System in the Field Radio Repairer Course.* U.S. Army Signal Center and Fort Gordon, Fort Gordon, GA, n.d.

King, F.J. *Evaluation of a Videodisc Delivery System for Teaching Students to Troubleshoot the AN/VRC-12 Medium-Powered Radio Series. Final Report.* Prepared for U.S. Army Training and Doctrine Command, Fort Monroe, VA, 1982.

Kirchner, G., Martyn, D., and Johnson, C. Simon Fraser University Videodisc Project: Part Two: Field Testing an Experimental Videodisc with Elementary School Children. *Videodisc/Videotex,* 1983, *3*(1), 45-57.

May, L.S. *Corporate Experience in Evaluating Interactive Video Information System Courses.* Paper prepared at Conference on Interactive Instruction Delivery, Orlando, FL (Society for Applied Learning Technology), February 1984.

Parker, J.E. A Statewide Computer Interactive Videodisc Learning System for Florida's CMS Nurses. *Computers in Nursing,* 1984, *1*(2), reprinted in *Computers in Nursing, 2*(2), 24-30.

Reeves, T.C. Alternative Research and Evaluation Designs for Intelligent Videodiscs. Paper presented at the American Educational Research Association, New Orleans, 1984.

Smith, R.C. *Full-Scale Pilot Testing of Florida's Videodisc Training Project.* Paper presented to the Conference on Interactive Instruction Delivery, Orlando, FL (Society for Applied Learning Technology), February 1984.

Thorkildsen, R. Interactive Videodisc for Special Education Technology. Final Report. Logan, UT: Utah State University, 1982. ERIC Document Reproduction Service No. ED 230 187.

Vernon, C.D. *Evaluation of Interactive Video Disc System for Training the Operation of the DCT-9000 in the MOS 72G Course. Final Report.* Prepared for U.S. Army Communicative Technology Office, Fort Eustis, VA, and U.S. Army Signal School, Fort Grodon, GA, 1984.

Wager, W. *CAI Evaluation for Basic Typing in MOS 72G10 Communications Specialist Course.* Prepared for U.S. Army Communicative Technology Office, Fort Eustis, VA, and U.S. Army Signal School, Fort Fordon, GA, 1984.

Wilkinson, G.L. *An Evaluation of Equipment-Independent Maintenance Training in the Programming of the AN/GSC-24 Multiplexer by Means of a Microprocessor-Controlled Videodisc Delivery System. Final Report.* The U.S. Army Training Developments Institute, Fort Monroe, VA, 1982.

Wooldridge, D., and Dargan, T. Linear vs. Interactive Videotape Training. *International Television,* August 1983, 56-60.

Yeany, R.H. *et al. Interactive Instructional Videotapes, Scholastic Aptitude, Cognitive Development and Focus of Control as Variables Influencing Science Achievement,* 1980. ERIC Document Reproduction Service No. ED 187 532.

Young, J.I., and Tosti, D.T. The Use of Videodisc and Microprocessor in Equipment Simulation. In *Equipment-Independent Training Program: Final Report.* Prepared for U.S. Army Training and Doctrine Command, Fort Monroe, VA, 1981.

A Comparison of the Effectiveness of Interactive Laser Disc and Classroom Video Tape for Safety Instruction of General Motors Workers

James Bosco and Jerry Wagner

In 1985, the United Auto Workers and General Motors initiated efforts to improve the instruction of workers on the handling of hazardous materials. The directive from the UAW-GM National Joint Committee on Health & Safety was to develop an instructional program which could be implemented in GM plants throughout the United States. The program was intended to accomplish two objectives: (1) to develop an effective training program for the handling of hazardous materials which would span the diverse operations of workers in General Motors, and (2) to deliver the training in a manner which insured acceptance of the training by the workers.

A committee to develop the training was established in 1985. As efforts to develop the training proceeded, it was decided to produce two versions of the training. One of the versions made use of video tape in a classroom format. Nine video tapes were produced along with detailed training manuals to guide the implementation of the training. The second version was a level three interactive video training program. This program presents action video, animated sequences, and computer text and graphics in an integrated instructional program which is controlled by the worker using the system. Through use of the touch screen, the worker can answer questions, choose topics to be presented, repeat or skip material, and obtain definitions of terms.

Many persons have called for objective evaluation of interactive video, since carefully designed data based evaluations provide a more secure basis for decision-making on the use of interactive video than do testimonials from those who develop interactive video. Previous discussions of evaluations have identified numerous problems in the design of evaluations (Bosco, 1986; Brody, 1984; Reeves, 1984). One of the major problems is the attempt to contrive an experimental model by finding a "control" treatment as a basis for the comparison of the "experimental" treatment. In this study, a natural comparison was possible because two parallel treatments were simultaneously developed. These two treatments were intended to cover the same content with two different delivery systems. This unusual situation offered a special opportunity in the development of the evaluation.

One of the important issues in this evaluation was the attitude of GM personnel toward the instruction. In designing both the video disc and video tape instruction, efforts were made to create programs which workers would find relevant and interesting. In evaluations of this nature, it is common to expose persons to one type of instruction and to assess attitude toward this type of instruction in isolation or by asking them to compare it to some other type of instruction they have experienced in the past (often "traditional" or "conventional" instruction). In either case, the comparative referent is unclear. In this study, the availability of two versions of the same instruction provided an opportunity to assess worker preference for the training approaches after they had direct experience with each of them.

The purpose of the evaluation was to assess the effectiveness of the Interactive Laser Disc System (ILDS) Training Program in comparison with the classroom instruction with video tape (Tape). The evaluation was focused on six major questions: Do workers who use the ILDS learn more than those who use the classroom training with Tape? What is the attitude of workers toward each of the two methods? What differences are there between the two methods with regard to amount of time required for the training? Which method do workers prefer? To what extent are worker age, sex, years of experience at GM, and years of education factors affecting the extent of learning produced by the ILDS? To what extent are worker age, sex, years of experience at GM, and education factors affecting the attitude toward the ILDS?

Procedure

Subjects

The subjects for the evaluation consisted of 209 workers from 15 GM plants in the Mid-West. The

James Bosco is Director, Merze Tate Center for Research and Information Processing, Western Michigan University, Kalamazoo, Michigan. Jerry Wagner is Training Coordinator of the UAW-GM National Joint Committee on Health and Safety.

criterion for recruitment of the subjects was that they could not have previously received training from the module chosen for use in the experiment, i.e., "Solvents."

General Design of the Study

Workers came to the UAW-GM Human Resource Health and Safety Training Center in Madison Heights, Michigan for a one-day training session. Each worker was randomly assigned to receive either the ILDS or Tape instruction on the training modules, "Solvents." Following this training, the worker completed a test of achievement on the module and an opinion survey about the instruction. Also, workers who had received the ILDS training were given a scale to measure attitude toward the ILDS as an instructional method. Each worker was then given training using the other method of instruction. This was done in order to provide the subjects with an immediate basis for comparison and to enable the subjects to report on their preference with regard to the two methods. While it is common to administer only one type of training and to solicit information from subjects regarding their preference (i.e., Do you prefer this type of instruction or "conventional instruction?"), it is often unclear in such cases what is really being compared. To eliminate this problem and to make the basis of the comparison explicit, workers were trained with each type of instruction. Following the training with the second type of instruction, each worker was administered two instruments: an opinion survey about the instruction they had just received and a user preference instrument. Additionally, if the ILDS was the second type of training they received, they also completed the attitude scale. After this testing was completed, subjects were interviewed. The testing procedures are summarized below:

1. Presentation of training—either ILDS or Tape
2. Achievement test
3. Opinion survey
4. Attitude scale (if subject had received ILDS training)
5. Presentation of second type of training—either ILDS or Tape depending on which training subject received first
6. Attitude questionnaire
7. Attitude scale (if subject had received ILDS training)
8. Debriefing interview

Logistics of the Training

Workers came to the UAW-GM Human Resource Training Center in 11 groups ranging in size from 10 to 28. The sessions began at 9:00 AM with a briefing about the study. At this time they received an explanation of the nature of the study and the agenda for the session. An important purpose of this briefing was to minimize any concerns about the testing. The key point made during this presentation was that the purpose of the study was to evaluate the methods of instruction—not them.

Two approaches were used for administering the ILDS and Tape training. Some groups were divided into two sub-groups at random, with each sub-group receiving one type of instruction (i.e., one sub-group went to the ILDS training while the other went to the Tape training). On other days the entire group was tested with one type and then administered the second type. Care was taken to balance the assignment of subjects with types of instruction so that an equal number received each type of instruction first. Ninety-five subjects received training using the first method of administering the two types of training, and 114 subjects received the second method.

The Nature of the Training

In a comparative evaluation such as this, clarity about the nature of each of the treatments being compared is critical. The ILDS was administered by having each subject sit at a computer station and work independently at the training. A trainer was present to answer any questions about the format, and a technician was available to debug any mechanical problems that arose, such as problems with the touch screen, etc.

A fair comparison required that the Tape training be handled in a way compatible with effective training procedures. The desired comparison was not between good training and poor training, but between two types of well designed training. Care was taken to find a trainer with a reputation for being effective. Another problem in the administration of the training was seeing to it that all workers received the same treatment. This was not difficult in the case of the ILDS but presented more of a problem when a trainer is involved. It is very difficult for a good trainer not to make corrections and modifications in the training as a result of experiences in earlier sessions. The need to maintain consistent functioning was explained to the trainer, and observers monitored the training to ensure that such occurred.

Pilot Test

Prior to beginning the evaluation, a pilot test was conducted. This consisted of a complete run-through, with the evaluation project personnel, of all of the procedures for the study. Nineteen workers participated in the pilot study. This pilot study enabled a number of logistical problems to be eliminated.

Instruments

1. **Achievement test.** The achievement test was constructed by first producing a content guide of the instruction in the solvent module. This was done to ensure coverage of the objective of the module. Next, a pool of questions was generated. These questions were then reviewed to judge their saliency. After the review, questions were modified. They were then reviewed by content experts to verify the correctness of the answers. They were then administered to a group of twenty subjects to compute reliability coefficient using Cronbach's Alpha. They were found to have a reliability coefficient of .67. A second Alpha was run using the data from the full scale study. The Alpha generated in this run was .66.

2. **Attitude scale.** The attitude scale was a revision of a scale developed by Finch (1969) called "An instrument to assess student attitude toward instruction."[1] This scale has a reported Kuder-Richardson reliability of .91. The Cronbach Alpha computed for the pilot study subject testing was .95. The reliability coefficient of the revised scale was found to be .92.

3. **Opinion survey.** An opinion survey was developed in order to obtain information from subjects about each type of instruction. When used in conjunction with the ILDS, an additional question pertaining to the ease of using the equipment was included.

4. **User preference.** This was a five-item questionnaire which obtained information from the workers about their preference with regard to the approach to training.

5. **Interview.** The interview contained four open-ended questions.

Findings

Workers in the evaluation ranged in age from 22 to 66 with a mean age of 41.3 years. The number of years employed at GM ranged from 2.3 to 40 with a mean of 14.4 years. The years of education of those participating in the study ranged from 8 to 16 with a mean of 12.6 years of schooling. There was considerable variability among the workers with regard to age (standard deviation = 7.6) and years of experience at GM (standard deviation = 8.1), but less variability with regard to education (standard deviation = 1.7). There were 156 males and 53 females in the sample.

Question 1: Was there a difference in achievement resulting from the use of the two different methods?

Table 1

Means on Achievement Test and Significance of Differences on Post-test Scores Between Workers Receiving Tape or ILDS Training

	Tape (n=104)	ILDS (n=105)	F and t Values
Mean	5.67	3.01	
Standard deviation	2.42	2.01	
F value			1.45
p for F value			.06
Pooled t			8.60
p for t value (df=207)			< .001

The first question for the evaluation pertained to the differences in achievement resulting from training with the two different methods. Table 1 provides a summary of the analysis of the differences in achievement resulting from training.

The Tape group had a mean number of errors which was 2.7 higher than the ILDS group. An F ratio was computed to examine the differences in variability of the two groups, and this F was not significant. Therefore, a pooled t test was used. This resulted in a significant t with a probability of less than .001. Thus, there is a statistically significant difference between the two types of training with a higher achievement resulting from the ILDS group.

Another way to assess the difference between the two types of training, is to examine the high and low achievers in the test. The mean for the entire 209 subjects as one group was 4.3 and the standard deviation was 2.6. More than one standard deviation above or below the mean represents the more extreme scores, i.e., the higher or lower achievers. There were 62 high achievers, i.e., persons with low number of errors and 41 low achievers, i.e., persons with high number of errors. Figure 1 shows the percentage of the high and low achievers who were trained on the ILDS or Tape method.

Figure 1 shows that the preponderance of the high achievers were trained with the ILDS training, and the preponderance of low achievers were trained with the Tape method.

One other analysis was performed to determine if the approach to administering the training in the evaluation (either by splitting the group or by presenting training sequentially to the entire group)

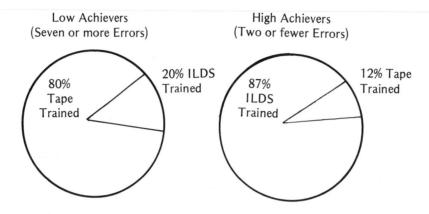

80%
Tape
Trained

20% ILDS
Trained

87%
ILDS
Trained

12% Tape
Trained

Figure 1. Percentage of total group of high and low achievers and the type of training they received.

affected the outcome. An examination of means from groups broken according to the two methods indicated that the approach used to administer training for the sample did not affect the results.

Question 2: What is the attitude of workers toward each of the two methods?

Data in response to this question were collected from two instruments. One of the instruments was an opinion survey containing items pertaining to feelings about aspects of the training was used. Also, an attitude scale was administered to each worker following ILDS training. The responses to the opinion survey are summarized in Table 2.

There were generally more favorable responses to the ILDS than to the Tape type training. Substantially higher percentages of individuals (22%) felt that the ILDS training would help them make safe use of solvents. There was also a substantially higher percentage (14% higher) of favorable response to the ILDS on the question about how interesting the training was. It should be noted, however, that there was only a small difference between the two groups with regard to their findings about pacing of instruction. Most of each group found the pacing to be "just right," and there was only a 4% difference between the two groups. Information from the interviews (discussed in the next section) suggests that workers may not have understood this question.

An attitude scale was administered to all subjects following the completion of the ILDS training. The highest possible score on this scale was 100. The lowest possible score was 20. The mean for workers on this scale was 77.1 (standard deviation 11.3). This indicates that workers tended to re-

spond in a favorable manner to the items on the scale.

Question 3: Is there a difference in the amount of training time for the Tape and the ILDS group?

The mean amount of time for training using the ILDS and the Tape were quite similar. The mean amount of time for workers in the ILDS was 33.87 minutes. There was a wide range for the amount of time taken by workers using the ILDS. The range was from 20 minutes to 74 minutes. The standard deviation was 7.70 minutes. The wide range of time taken by workers using the ILDS indicates that workers did avail themselves of the opportunity to move through the training at differing paces.

Question 4: Which method do workers prefer?

Data for this question were provided by a questionnaire which was taken by workers after they had experienced each type of instruction, and also through use of an open-ended interview. Approximately 80% of the workers indicated a preference for the ILDS, and 75% of the subjects indicated that they felt the ILDS would be preferred by other GM employees. Table 3 presents a summary of responses to the user preference text.

Consistent with the data from the attitude scale and the opinion survey, the interviews reflected more of a positive response to the ILDS than to the Tape. There were several recurring themes in the interviews. One of the things workers tended to single out for comment was that the ILDS method provided an opportunity to review material they did not understand. Along the same line, a number of subjects felt that the instruction on the video tape moved too quickly. Several subjects

Table 2

Responses to Opinion Survey
with Reference to the Tape and ILDS Training

Item	Response	Tape f (n=209)	%	ILDS f (n=208)*	%
Help safe use of solvents	A lot	124	59	169	81
	Some	81	39	38	18
	Little	4	2	1	1
Training pace	Too fast	31	15	18	9
	Just right	174	83	181	87
	Too slow	4	2	7	3
	Other	-	-	1	1
How interesting?	Very interesting	102	49	136	65
	OK	98	47	69	33
	Not too interesting	9	4	3	1
Recommend training to others?	Yes	173	83	193	92
	No	33	16	13	6
	Unsure or other	3	1	2	1
Change way you use solvents?	Yes	167	80	190	91
	No	41	20	18	9
	Unsure or other	1	1	1	1
Opinion of training	Very good	78	37	122	59
	Good	83	40	62	30
	OK	40	19	17	8
	Poor	7	3	6	3
	Very poor	1	1	1	1
	Other	-	-	1	1
Ease of using ILDS	Easy	-	-	158	76
	Little difficult	-	-	47	23
	Very difficult	-	-	3	1
	Other	-	-	1	1

(n=209)
*Note: One opinion survey was missing for ILDS group.

felt that the video was boring or sleep provoking. The most common concern or negative comment about ILDS was that it did not provide an opportunity for interaction. Some felt better about the Tape since there was discussion and group interaction.

Questions 5 and 6: To what extent are worker age, sex, years of employment at GM, and years of education factors affecting the post-test achieve- ment scores and the attitude scores for workers in the ILDS group?

The results on the achievement test were examined in relationships to four potential explanatory variables. Table 4 contains a correlation matrix for the four variables. Apart from the correlation of age with years of employment, the correlations among the four variables were all not significant. The only significant correlation among

Table 3

Summary of Responses of Workers About Training with Regard to Aspects of Training

	Prefer Tape		Prefer ILDS		No Preference		Missing Data
	f	%	f	%	f	%	f
Which training:							
Easier to pay attention to?	42	21	157	77	5	3	5
Best training pace?	36	18	163	80	5	3	5
Produced more learning?	42	21	160	78	2	1	5
Preferred?	37	18	164	80	3	1	5
Likely preference by other GM employees?	46	23	155	76	3	1	5

(n=209)

Table 4

Correlation Matrix for Test Scores, Years of Employment at GM, Age and Years of Education

	Test Score	Years of Employment	Age	Years of Education	Sex
Test Score	1.00	-.02	.19	-.40	.16
Years of employment	-.02	1.00	.41	.06	.10
Age	.19	.41	1.00	-.16	-.10
Education	-.40	.06	-.16	1.00	-.24
Sex	.16	-.10	-.10	-.24	1.00

(n=99*)
*Note: This number varies from the n for this group because cases with missing data are excluded.

Figure 5

Correlation Matrix for Attitude Toward ILDS, Years of Employment at GM, Age and Years of Education

	Attitude	Years of Employment	Age	Years of Education	Sex
Attitude	1.00	-.02	-.17	-.15	.14
Years of employment	-.02	1.00	.41	.08	-.10
Age	-.17	.41	1.00	-.15	-.12
Education	-.15	.08	-.15	1.00	-.23
Sex	.14	-.10	-.12	-.23	1.00

(n=95*)
*This number varies from the n for this group because cases with missing data are excluded.

these variables was between test score and years of education. Higher scores were somewhat associated with more years of education.

A multiple regression analysis was performed in order to examine the extent to which these four variables taken together explain test performance. The multiple R for these four variables was .44. This means that these four factors together explain only about 16% of the total variance of test performance. In other words, the four characteristics investigated (age, sex, years of employment at GM, and years of education) have little bearing on test performance. What little bearing they do have is almost entirely that of years of education.

The scores on the attitude survey were examined in relationship to the four explanatory variables. The relationship between the four explanatory variables and attitudes was weak. The variables had little bearing on attitude toward ILDS. (See Table 5 for the correlation matrix for these variables.) The multiple regression coefficient for these variables was .27. This means that these four variables only account for 7% of the variation in scores on the attitude scale.

One of the major problems which has afflicted evaluations of interactive video has been the faulty design of "control group" treatments. The interactive video treatment is often compared with the instruction that interactive video is meant to replace. In such cases, the Hawthorne Effect is a likely prospect, and the results are biased in favor of the new method. In this evaluation, the development of two new parallel presentations of the same content provided an unusually good opportunity to run a comparative test and to control for the Hawthorne Effect.

The two basic questions which were the focus of the evaluation were: Does the ILDS result in the effective achievement of training information by workers? and, How do workers feel about the use of ILDS? The results of this evaluation produced positive findings for the ILDS on both of these questions. Workers who used ILDS learned more than those who used the Tape type training. The Tape group had more mean errors on the post-test and also produced a higher standard deviation. The ultimate goal in training is to achieve a zero mean number of errors and a standard deviation of zero on the post-test. This would indicate that all workers had achieved a complete grasp of the training information. The ILDS group came closer to this goal than did the Tape group.

Worker attitude toward the use of the ILDS was highly positive. Since workers were given an opportunity to experience both types of training, the data on worker preference for the type of training take on added importance. When persons are presented with an instructional treatment and asked how they like it, they tend to respond favorably. The data on attitude take on more value when persons can compare two or more types of instruction which they have experienced. In this evaluation, workers were given the opportunity to report on their attitude toward interactive video—not in an abstract sense, but in comparison with another training procedure which was designed to be effective and attractive.

The data on worker attitude indicated clear preference for the ILDS. Workers tended to speak about the advantage of ILDS in terms of the opportunity to move at their own pace and the way the ILDS reduced boredom and distraction. The number of workers with a strong negative opinion of ILDS was small (less than 5%), but a larger group (about 20%) did indicate a preference for Tape type training (without strong negative feeling about ILDS training). The most common concern expressed by workers about the ILDS was the lack of human interaction. Some, even some of those who were in favor of ILDS, expressed concern about losing the opportunity to get training in groups so that interaction among the group is possible.

Worker success on the post-test and attitude toward the instruction was not a function of age, years of employment at GM, or sex of worker. There is some effect of years of education on post-test success, but even this variable is not strongly related. Years of education is not related to attitude toward the instruction. As noted, these background characteristics had little impact on either type of training.

In summary, the evaluation strongly suggests that the interactive video instruction was effective in achieving the training outcomes of the module used for the evaluation, and workers liked it. There is one important caution in interpreting these findings. This evaluation was conducted in a "laboratory" setting. Even though this evaluation indicates that ILDS can work, the manner in which it is implemented in the work environment is critical to its success as an instructional tool. Care must be taken to introduce, maintain and support the ILDS in order for it to produce the intended training outcomes reflected in this evaluation. □

References

Bosco, J. An Analysis of Evaluations of Interactive Video. *Educational Technology*, 1986, *26*(5), 7-17.

Brody, P.J. *Research on and Research with Interactive Video.* Paper prepared for the annual meeting of the

American Educational Research Association, New Orleans, 1984.

Finch, C.R. An Instrument to Assess Student Attitude Toward Instruction. *Journal of Educational Measurement*, 1969, 6(4), 257-258.

Reeves, T.C. *Alternative Research and Evaluation Designs for Intelligent Videodiscs.* Paper presented at the Ameri-

can Educational Research Association, New Orleans, 1984.

Note

1. Appreciation is acknowledged to Dr. Curtis Finch for permission to use his instrument in this study.

How Effective Is Interactive Video in Improving Performance and Attitude?

David W. Dalton

Introduction

Computer-assisted instruction (CAI) has had beneficial effects on learner achievement in a wide variety of instructional settings. Research has shown that CAI not only improves learner achievement, by as much as 50 percent, but that it can also reduce the amount of time necessary to accomplish the same amount of learning (Kulik, 1983).

CAI has been effective with a wide variety of learners and in many different types of instructional settings (Charp, 1981). In addition, CAI has had positive effects on improving the affective outcomes of instruction, such as learner attitude and self-esteem (Dalton and Hannafin, 1985; Clement, 1981). The favorable attitudes of learners who participate in computer-assisted instructional programs have been attributed to the fact that the computer had infinite patience, never showed signs of anger or frustration, and left the learners with a general feeling of having learned "better."

Yet, despite the many instructional benefits associated with the use of CAI, there are many instructional situations in which CAI simply is not adequate or appropriate (Martorella, 1983). For example, computer generated graphics are generally not capable of depicting intricate, visually-oriented instructional sequences, such as surgical procedures or flight training, with the realism that is required.

On the other hand, video images can present instruction with a realism that is not possible in CAI. However, although video-based instruction has been effective in many situations, the many instructional benefits of typical CAI are lost (Russell, 1984). Since video-based instruction is generally non-interactive, the possibilities for individualized pacing, feedback, and reinforcement are greatly diminished.

David W. Dalton is Assistant Professor of Education, Instructional Systems Technology, Indiana University, Bloomington, Indiana.

Many authors note that video often becomes a passive instructional medium where learners do not actively participate in the learning and hence simply "turn-off" to the instruction (Gendele and Gendele, 1984).

In the past decade, computer and video technologies have been merged to form a medium known as "interactive video." This encompasses interactive videotape and interactive videodisc technologies. In the research reported in this article, a *tape* player was employed. With interactive video technology, the learners are shown a segment of video instruction and asked questions about that segment by the computer. The computer can then perform the same functions as it does in more conventional CAI: inputting and judging the learners' responses, providing feedback and reinforcement, and recordkeeping.

The possibilities for improving CAI with video images through interactive video instruction seem very promising. Current research indicates that the variety of visual and auditory learning stimuli present in interactive video can dramatically improve learning (Clark, 1984).

In addition, a recent study noted that the interactive nature of interactive video can not only improve short-range recall, but can also aid in retention (Schaffer and Hannafin, in press). However, this study also demonstrated that *excessive* amounts of interactivity in interactive video do not appreciably affect performance or retention, but drastically impact the efficiency of the instruction presented.

Although assumptions made about interactive video make it seem ideally suited for many educational and training settings, there are questions concerning the use of interactive video technology that have yet to be answered. The study reported in this article compared the effects of interactive video instruction on learner performance and attitude, with conventional CAI and stand-alone video, in order to determine exactly what types of learning tasks best lend themselves to interactive video instruction.

Materials and Methods

The 134 subjects for this study were selected from six introductory level junior high Industrial Arts Exploration classes. The basic learning consisted of a set of 27 General Shop Safety Rules. Each of the rules involved a visually-oriented task or behavior required of the learners. Three parallel forms of instruction were employed: Video-only, CAI only, and Interactive Video. The video-only lesson consisted of a 15-minute video presentation on the safe use of tools. Learners were shown a short, narrated segment that depicted both an example and non-example of the correct behavior. The CAI-only lesson used the narrator's script as the basis of a tutorial lesson. The interactive video lesson combined the video segments from the video lesson with the tutorial from the CAI lesson.

Prior to the beginning of the study, the learners were designated as relatively high or low in prior achievement based on their sixth grade total *Comprehensive Test of Basic Skills* scores. They were then randomly assigned to one of the three treatment groups described. At the conclusion of the lesson, the learners were given a print-based posttest, covering the rules that had been presented in the three treatments, and a survey to address their attitudes towards the instruction.

Design and Procedures

This study employed a completely crossed 3 x 2 x 2 treatment by achievement by sex factorial design, featuring three levels of treatment (video only, CAI only, and interactive video), and three levels of prior achievement (high, average, and low) based on CTBS scores. Dependent measures included one measure of performance and one measure of attitude toward instruction.

Posttest performance scores were analyzed with ANCOVA procedures, using prior achievement as the covariate. Attitude scores were analyzed with ANOVA procedures.

Findings

The means for the treatment groups on the performance measure were 64.98 percent, 73.54 percent, and 70.48 percent for the Video, CAI, and Interactive Video treatments, respectively. The means of all three groups were significantly different at the alpha = .05 level.

In order, the attitude scale means of the Video, CAI, and Interactive Video treatment groups were 75.07 percent, 74.26 percent, and 82.87 percent. The mean of the Interactive Video group was significantly higher than both the CAI and Video only groups at the alpha = .005 level. However, the means of the Video group and the CAI group were not statistically different.

In addition to the treatment main effect on the attitude scale, there was also a significant Achievement by Treatment Interaction.

Implications of the Study

There are three major findings from this study that warrant discussion: (a) CAI alone tends to the the most effective instructional delivery system for the type of learning task chosen for this study, (b) interactive video instruction produced significant improvements in learner *attitudes* when compared with CAI and Video alone, and (c) the attitude effects observed in this study were not constant across prior achievement level.

It might be assumed that the interactive video treatment, by virtue of its video enhancements *and* individualization, would be the most effective in producing high levels of performance. However, this assumption was not supported by this study for two principal reasons. First, the interactive video equipment was very new to these students. Observations made during the lessons indicated that the learners were somewhat distracted by the various noises and indicator lights produced by the videotape players in this treatment. Second, the delays caused by long tape access times* may have given the learners the opportunity to drift and not actively participate in the instruction. On the other hand, these learners were familiar with CAI lessons, so this medium provided no such distractions and its more direct nature seemed to keep these learners more "on task."

Although the learners using the CAI lesson performed best, the interactive video lesson was successful in improving learner attitudes towards the instruction. This improvement in learner attitude may be the result of the more motivating nature of the "natural" video images or the immediate reinforcement provided by the computer (Bejar, 1982). Unfortunately, as noted earlier, the learners involved in this study had never used this kind of delivery system before. Therefore, the differences in observed attitudes may, in part, be attributable to a novelty effect.

However, the most important finding of this study is that the attitude differences observed varied across prior achievement level. Specifically, low ability students scored disproportionately lower on the CAI lesson than low learners in the other treatment groups. What, then, could account for this strongly negative reaction to the CAI treatment by low ability learners?

In the school chosen to participate in this study, CAI had been used for approximately four years, primarily for remediating the basic skills deficiencies of low-ability learners. Perhaps CAI, when used only in a remedial capacity, can have the same stigmatizing effects often observed when low ability learners are placed in conventional "special" programs. The results of this study support the notion that a great deal of care is warranted in the use of CAI, and remedial programs in general, if these detrimental effect are to be avoided.

In summary, the results of this study indicate that CAI can be a highly effective mode of

*The faster access times offered by video*disc* technology conceivably might alter the results obtained.

instruction where the additional capabilities provided by interactive video are not required. In addition, both interactive video and CAI are more effective in producing high levels of performance than video only, substantially due to their ability to keep learners more actively participating in the learning. Finally, although CAI can be used to effectively improve learner attitudes, like other types of instructional media, CAI can have deleterious effects on learner attitudes if used in a manner where low ability learners feel demeaned or isolated because of their additional needs. □

References

Bejar, I. Videodiscs in Education: Integrating the Computer with Communication Technologies. *Byte*, 1982, *7*(1), 78-104.

Charp, S. Effectiveness of Computers in Instruction. *Viewpoints in Teaching and Learning*, 1981, *57*(2), 28-32.

Clark, J. How Do Interactive Videodiscs Rate Against Other Media? *Instructional Innovator*, 1984, *29*(6), 12-16.

Clement, F. Affective Considerations in Computer-Based Education. *Educational Technology*, 1981, *21*(4), 28-32.

Dalton, D., and Hannafin, M. The Effects of Computer-Assisted Instruction on the Self-Esteem and Achievement of Remedial Junior High School Students: An Exploratory Study. *Association for Educational Data Systems Journal*, 1985, *18*(3), 172-182.

Gendele, J.F., and Gendele, J.G. Interactive Videodisc and Its Implications in Education. *Technological Horizons in Education*, 1984, *12*(1), 93-97.

Kulik, J. Effects of Computer-Based Teaching on Secondary School Students. *Journal of Educational Psychology*, 1983, *75*(1), 19-26.

Martorella, P. Interactive Video Systems in the Classroom. *Social Education*, 1983, *47*(5), 325-7.

Russell, A. From Videotape to Videodisc: From Passive to Active Instruction. *Journal of Chemical Education*, 1984, *61*(10), 866-68.

Schaffer, L., and Hannafin, M. The Effects of Progressive Interactivity on Learning from Interactive Video. *Educational Technology and Communication Journal*, in press.

The Validation of an Interactive Videodisc as an Alternative to Traditional Teaching Techniques: Auscultation of the Heart

Charles E. Branch, Bruce R. Ledford,
B.T. Robertson and Lloyd Robison

Introduction

Biomedical education, especially physiology, traditionally has used live animal experimentation to teach biological principles and problem-solving skills. Many educators in the biomedical sciences believe that such experiments are essential in introducing students to the special nature of living organisms. Most biomedical educators would at least agree that such experiments can be very effective in teaching some areas which are difficult to teach using the traditional lecture format.

The arguments over effectiveness of animal experimentation may be academic. Such experiments are under pressure from both without and within the traditional biomedical education establishment. Effectiveness may not be the only, or even the most important, criterion used to judge experiments in the future. In some countries animal experimentation has been eliminated from the curriculum except in very specific and limited circumstances.[1]

The trend has also been evident in the United States. In some teaching situations computer-assisted instruction, computer simulations,[2] or videotape lessons have been developed as effective alternatives. Unfortunately, in many other instances, effective laboratory exercises have simply been deleted from the curriculum with no substitution of alternatives. Many educators believe that this trend will produce students deficient in the basic understanding of biological principles and essential problem-solving skills.

For some time we have been interested in the development of interactive videodisc and videotape as alternatives to traditional laboratory exercises.[3] Interactive video consists of a videodisc or videotape player under control of a computer program. The student interacts with the computer program, so that information from the video player and/or the computer is presented based on individual responses of the student. Ideally the student is insulated from the hardware, and is not even conscious of the source of the information presented.

A single videodisc can contain various combinations of up to 54,000 single pictures or up to 30 minutes of video action. Unlike the linear videotape process, individual frames or segments can be located very quickly. Thirty minutes of audio can also be included on each of two separate audio tracks. Many hours of additional audio information can be included by use of still-frame audio.

Interactive video can combine the advantages of traditional computer-assisted instruction (CAI) and videotape instruction. Like videotape instruction, the information presented to the student can be very realistic. As in traditional CAI, presentation of the material can be adjusted to the specific needs of individual students. This merging of two related technologies could provide students with very realistic simulations of actual experiments. The "experimental" results could vary depending on responses of the student, permitting the student to test various procedures, see the actual results, and even test procedures not possible when using live animals.

Previous Methods

The videodisc used in this study is on heart sounds and murmurs. Auscultation, the process of analyzing sounds associated with cardiovascular function, is largely an art. Developing skill requires repetitive comparisons between various normal and abnormal sounds supported by a firm understanding of the cardiac cycle. This ultimately requires years of experience auscultating patients, and no use of technology will substitute for this experience.

Previous methods of teaching this material have forced instructors to maintain colonies of abnormal animals, to use surgically prepared experimental animals, or to rely on animals with abnormal heart sounds which happened to be available in the clinic when the course is taught. The animals were brought to the classroom, and the heart sounds were played for the students using an

Charles E. Branch and B.T. Robertson are with the Department of Physiology and Pharmacology, College of Veterinary Medicine, and Bruce R. Ledford and Lloyd Robison are with the Department of Educational Media, College of Education, Auburn University, Auburn, Alabama. The research reported in this article is supported by grants from the Merck Company Foundation and the Geraldine R. Dodge Foundation.

electronic stethoscope, amplifier, and speakers. This was a difficult and time-consuming process, and occasionally it was totally unsuccessful. The availability of audiotapes and records[4] has helped considerably, but these require the students to spend time dealing with the hardware itself. Rapid comparison of similar sounds resulting from different abnormal conditions is difficult. Furthermore, records and tapes use only the sense of hearing. They give little help in relating the cardiac cycle to the audio sounds.

The major difficulty is that most students are disoriented for some time after initial exposure to auscultation. They have considerable difficulty relating the unfamiliar sounds to the cardiac cycle. Substantial time is wasted before they become oriented enough to begin developing skills.

To help students through this transition period we have supplemented the records and audiotapes by using an oscilloscope to display the video depiction of the audio sounds. This was useful in the formal classroom or laboratory, but operation of the oscilloscope was sufficiently complex that it was not practical for use by individual students or groups of students. We decided that we needed a way to provide a variety of audio recordings supplemented with the visual senses, but not requiring the student to deal with the hardware. The videodisc was developed to meet that need.

Statement of Purpose

The purpose of this study was to validate the effectiveness and student acceptance of an interactive videodisc as an alternative to traditional techniques in teaching auscultation of the canine heart.

Hardware System

The overall hardware system used in the project has been described.[5],[6] The system used in the current study (Figure 1) consisted of an IBM PC computer, Pioneer LD-V6000 Videodisc player, Sony KX-1211HG RGB/Composite monitor, and a system designed in-house to overlay computer generated text onto the video image from the video player. The video player was controlled directly from the RS 232 serial port of the computer.

Software

The software consisted of the video/audio recordings and the computer program. The techniques for generating the recordings and converting them to videodisc have been described.[7] In summary, the heart sounds were recorded from an electronic stethoscope onto one audio channel of a high fidelity video recorder. The electrocardiogram

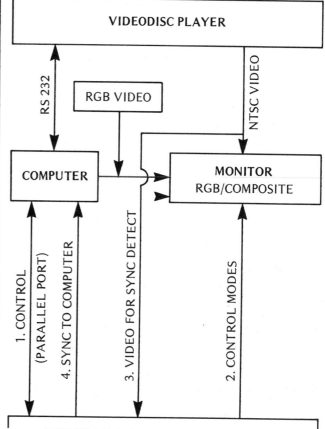

Figure 1

System Used in Study

Block Diagram of Student Videodisc System. The computer communicates with the videodisc player through the RS 232 serial port. The overlay circuit detects the video synchronization pulses from the composite video signal and then synchronizes the computer's RGB video with the composite video. The mode of operation of the monitor is controlled from the parallel printer port of the computer.

(EKG) was recorded on the second audio channel using an FM adapter. The recordings were played back onto a storage oscilloscope to give a continuous display of the EKG and a video depiction of the audio heart sounds. Other existing heart sound recordings without EKG were also included. The video image from the oscilloscope was recorded on a videotape recorder along with the original audio.

The videotapes were edited, and a second audio channel was dubbed in to provide voice descriptions of the cases. Still-pictures of selected diagrams, necropsy slides, and other data were added during editing. The videotape was converted to a one-inch master, and a video check-disc was made by a commercial video lab (Spectra Image, Burbank, CA). The videodisc contains 35 sequences of about 40 seconds each, in addition to several audio-only segments and about 60 still-pictures.

A computer program was written to interact with the user and control the videodisc player. The program presents some optional introductory screens, a main menu, and a series of sub-menus. The main menu consists of the major categories of recordings. Choosing a category results in the presentation of a sub-menu from which actual recordings are selected.

An additional feature was added to the program to facilitate comparison of different recordings. From the main menu the user may activate a "Toggle" mode, and then select any two recordings from any of the menus. Then the user may alternate quickly between the two recordings by pressing a single key.

The videodisc contains only the audio/video recordings and pictures. All menus and labels are generated from the computer and overlaid onto the video images. This gives higher quality text than can be displayed with composite video and also permits modification to the computer program without having to modify the videodisc.

Program Operation

The program initially presents a series of introductory screens from the computer. The user may read the screens or bypass them and go directly to the main menu. The main menu gives options for various categories of recordings, e.g., Normal Heart, Mitral Insufficiency, Congenital Defects, or Arrhythmias. The toggle mode described above may also be activated from the main menu. Selection of a category results in one of the sub-menus, for example, Normal Heart. From the Normal Heart sub-menu the user may select from a variety of normal recordings. Color codes indicate whether the user has already played a recording. After the selection of a recording a window menu permits the user to choose from various playback options, e.g., Recording Only, Description, or Still-frame.

For example, if the user selects option D (description) by pressing the "D" key, then the recording is played with the heart sounds coming from one speaker and the voice description coming from the other speaker. During the playback a dynamic window menu at the bottom of the screen permits the user to pause, change playback modes, go to a still-frame, or end and return to the sub-menu.

If the user, for example, chooses the still frame by pressing the "S" key, then the disc goes immediately to the still frame. Overlaid labels pointing out systole, diastole, or other features help the user relate the audio sounds to the cardiac cycle. At this point the user may return to the recording or end the sequence and return to the sub-menu.

The disc contains a total of 35 normal and abnormal recordings of hearts sounds, of which the above is one example. Other choices in the sub-menus permit the user to see pictures or diagrams related to the recordings, e.g., diagrams of the abnormal blood flow paths or photographs of cardiac lesions which could cause the abnormal heart sounds.

Methods

Subjects for the study were 87 first-year veterinary medical students who were enrolled in a cardiology course during the spring quarter, 1986, at Auburn University. The subjects were randomly assigned to either an experimental or control group. The learning activities were the controlled independent variables, while subject performance on a practical examination was the dependent variable; also, subjects' attitudes to several questions were compared using a survey.

Both groups attended classes in which the theory of auscultation was discussed and in which examples of heart sounds were played using available cases from the clinic. Both groups were given copies of a record of canine heart sounds,[8] and were advised to use the record on their own time and keep a log. In addition, audio cassette tapes were made from the videodisc master tape and were placed in the library for check-out.

During the first formal lab the control group was exposed to audio heart sound recordings in the same manner as done in the past. The experimental group was exposed to the videodisc, and samples of the sounds were played from the disc. Following the first lab the students practiced on their own, with the videodisc available to the experimental group only.

We observed that the students usually used the system in groups of two to six students. Frequently we observed one student operating the system, with the other students listening to the recordings, apparently engaged in self-testing.

Table 1

Comparison of Survey Results Between Control and Experimental Groups

Assignments of Values for Survey:
- 4 = Strongly Agree
- 3 = Agree Somewhat
- 2 = Disagree Somewhat
- 1 = Strongly Disagree
- – = No Opinion (Not included in data). The students were requested to choose this response only if they had no basis for an opinion or did not understand the question.
- N = Number answering question
- Mean = Mean of those answering using above values

	Control Mean (N)	Experimental Mean (N)	t-Score
Group A (questions on methods)			
1. I was given sufficient resources to learn and practice auscultation.	3.0 (41)	3.6 (40)*	–3.42
2. I believe that my group had the better learning experience.	2.2 (39)	3.1 (38)*	–5.53
3. Lecture material and the heart sound record were sufficient without a need for other material.	2.4 (42)	2.5 (39)	–0.73
4. New video programs in the future should be introduced to all students simultaneously instead of just half the class.	3.5 (41)	3.6 (39)	–0.66
5. New video programs in the future should be evaluated as this was done before they are incorporated into the curriculum.	2.6 (39)	2.3 (37)	1.29
6. I believe this evaluation was fair to all students.	3.0 (41)	3.1 (37)	–0.40
7. I believe my group had an advantage in developing skills.	2.2 (39)	2.8 (39)*	–2.85
8. The heart sound record issued to us was a good way to develop skills.	3.0 (43)	3.4 (40)*	–2.88
9. The audio tapes placed in the library were a good way to develop skills.	3.1 (36)	3.4 (29)	–1.57
10. The video disc was a good way to develop skills.	3.7 (22)	3.7 (40)	0.21
Group B (general questions on auscultation)			
1. I believe that skill in auscultation of heart sounds is important.	4.0 (44)	4.0 (39)	–0.48
2. I think it is good that we spent time learning to auscultate.	4.0 (44)	3.9 (39)	1.00
3. I believe it would be better to wait until we get into the clinics before we concentrate on auscultation.	1.3 (44)	1.7 (39)*	–2.39
4. A firm understanding of the cardiac cycle is important in auscultation.	3.8 (44)	3.8 (39)	–0.77
5. I have confidence in my *current* skills in auscultation.	2.6 (44)	2.8 (39)	–1.24
6. I have confidence in my *potential* skills in auscultation.	3.5 (44)	3.6 (39)	0.45

* = significant difference between control and experimental groups.

Survey Data Analysis

The questionnaire (see Tables 1 and 2) asked for responses to three sets of questions. (A fourth group of questions dealt with another subject and is not included here).

Both groups answered the first two sets on the questionnaire (Table 1). Of those items, the following exhibited significant differences in response between the control and experimental groups.

Question A1: **We had sufficient resources to work with.** Both groups rated the available resources as positive, but the experimental group was significantly more positive than was the control group. This could result from the control group being aware that the experimental group was having an additional experience. The important thing is that the experimental group was highly positive about the experience.

Question A2: **My group had a better learning experience.** The experimental group definitely scored more positive in the response to this question. The question is, was the experience truly more positive, or did the control students see themselves at a disadvantage and, therefore, score the item down?

Question A7: **My group had an advantage.** This is a parallel question to the previous question, and again the experimental group's response was significantly higher than that of the control group. Comparing the two questions indicates that the experimental group perceived a better learning experience, but some of them did not see it in terms of advantages or disadvantages. That is, the experimental group was more positive to Question A2 than A7.

Question A8: **The heart sound record was good.** It is interesting that the experimental group scored the record significantly higher than the control group. This could indicate that the control group felt deprived, or it could be that the experimental group found the record more helpful after first having been oriented to auscultation using the videodisc.

Question B3: **Wait until clinics for training.** The response to the question indicates that both groups rated an affirmative very low, but the experimental group appeared less negative. An apparent discrepancy exists between the subjects' response to this question and question B2: *It's good that we spent time learning to auscultate.* Upon examination of the raw data for the two questions it was found that six of the experimental group rated both questions B2 and B3 at values of 3 or 4, an inconsistent response. Only two of the control group made this apparent error. Further examination revealed that no students in either group rated question B3 higher than a value of 2

except for those who made this apparent error. This accounts for the higher standard deviation for the experimental group and probably accounts for the difference itself.

None of the other questions indicated any significant differences. However, the responses to several questions are of interest. Question A6 indicated that both groups of students thought the evaluation was fair to all students, although question A4 indicates that both groups would prefer to introduce new video programs to all students simultaneously. This apparent conflict in attitudes is supported by the students' ambivalent response to question A5.

Table 2

Response of Experimental Group to Question on Videodisc

		Mean (N)	Ranking
GROUP C (specific questions on videodisc)			
1.	Overall, the videodisc system was helpful to me.	3.4 (39)	8
2.	The videodisc system was easy to use.	3.8 (40)	2
3.	The system saved me time.	3.1 (38)	10
4.	The system helped me relate the cardiac cycle to the heart sounds.	3.4 (40)	7
5.	The system was reliable.	3.5 (39)	4
6.	The system was available at reasonable and sufficient times.	3.6 (40)	3
7.	The audio quality of the disc was good.	3.4 (40)	9
8.	The video quality of the disc was good.	3.5 (40)	5
9.	The video helped me at first, but I gradually began just listening, rather than watching the screen.	3.5 (39)	6
10.	The system should add a practice self testing mode.	3.8 (39)	1
11.	More information (explanations, discussion of the cardiac cycle, animation of the cardiac cycle, etc.) should be added.	2.6 (37)	11

The third group of questions (Table 2) asked for comments on the videodisc program from the experimental group Overall, the experimental group gave the system good marks for quality. Note, however, that the item that got the highest priority was that the program should have a self-test mode added to it. This is consistent with our earlier observations of user behavior.

Note also that the experimental group found the video helpful at first, but then gradually began just listening (Question C9). This could help explain the experimental group's greater acceptance of the audio record (Question A8), which was used by both groups.

Analysis of Data

A multiple analysis of variance was used to test for differences between test scores. The total time spent using the media systems was used as a covariate for the analysis.

The total time spent on media by the control group averaged 304.89 +/- 224 (S.D.) minutes, while the experimental group averaged 360.88 +/- 187 minutes. It should be noted that the experimental group was given strong encouragement to fill out the logs of videodisc use. The computer program asked the user to fill out the log slip at the beginning and end of the program. Keeping logs of individual use of the record was also encouraged, but not at the actual time of use. It is possible that because of the reinforcement, the experimental group was also more conscientious about keeping logs of the other media.

The two groups were tested for homogeneity using a pretest covering lecture material before the beginning of the experiment. As a result it was found that they did not vary significantly from the predicted population norm. (Cochran's C yielded a score of .63472, which has a probability of 1.000.) The two groups had mean scores of 81.32 +/- 16.8 (control) and 83.26 +/- 19.0 (experimental) on the heart sound recognition test.

Although the experimental group had a mean score 1.94 percentage points higher than the control group, the multivariate test for significant difference between the two group resulted in a significance value of .149, indicating that the difference between the means was not significant.

Wilks lambda	F	hyp. d.f.	error d.f.	sig.
.95517	1.948	2.00	83.00	.149

Conclusions

We found the use of interactive video to be effective as an alternative to traditional techniques in teaching auscultation of the canine heart. The students were very receptive to this technique.

We feel that interactive video will be appropriate for applications where it can substitute for animal experiments, especially in substituting for abnormal animals which may not be available when needed. Material requiring repetition to develop proficiency, especially when students must develop skill observing and discriminating between subtle variations, offers another appropriate application. Developing material with widespread applicability could reduce total costs by amortizing production costs over a large user base.

At Auburn University, we are investigating interactive video for teaching cardiovascular physiology, hemorrhagic shock, and other subjects. We hope to encourage other schools to develop a cooperative effort, such as a consortium, to develop and distribute lessons to a variety of schools. Such a cooperative effort would help reduce costs by increasing the possible user base. An additional factor would be in the evaluation of lessons prior to distribution. This might help substitute for the lack of a peer review method for such lessons. The lack of peer review similar to that provided by publication of research results has greatly inhibited the development of computer assisted instruction, especially by younger faculty.

Animals used in teaching laboratories demonstrate current information; no new information is provided. Substitution of appropriate alternatives for animals would not reduce teaching effectiveness. Indeed, the opposite may be true: alternatives in teaching may permit us to reinstate simulated versions of experiments which have been deleted from the curriculum because of reduced available time, increased costs, and pressure from society.

In many instances interactive video simulations could accomplish all the benefits which accrue from actual experiments. In addition, since time devoted to experimental preparation could be condensed, it is likely that the same material could be covered in less time for both the students and the faculty. Interactive video used in this manner would benefit students, faculty, and society in general. □

Notes

1. U.S. Congress, Office of Technology Assessment. *Alternatives to Animal Use in Research, Testing, and Education.* Washington, DC: U.S. Government Printing Office, OTA-BA-273, February 1986.

2. J.A. Michael. Computer-Simulated Physiology Experiments: Where Are We Coming From, and Where Might We Go? *The Physiologist*, 1984, *27*(6), 434-436.

3. C.E. Branch, B.T. Robertson, E.P. Smith, and J.T. Vaughan. Interactive Video in Veterinary Education: An Alternative to Live Animal Experimentation. In *Proceedings, 2nd Symposium of the American Veterinary Computer Society*, 1984, 219-222.

4. S. Ettinger. *Canine Heart Sounds*, ESCO Pharmaceutical Corp., Buena, New Jersey.

5. C.E. Branch. Interactive Video in Veterinary Education: Hardware Alternatives. In *Proceedings, 2nd Symposium of the American Veterinary Computer Society*, 1984, 223-226.

6. C.E. Branch. Text Over Video. *PC World*, December 1983, 202-210.

7. C.E. Branch, and B.T. Robertson. Interactive Video Disc Simulated Animal Experiments. *J. Vet Med Education*, in press, 1986.

8. Ettinger, *op. cit.*

9. Michael, *op. cit.*

Instructional Design

Interactive Video: A Formative Evaluation

Roberts A. Braden

In our office for the past year or so we have spent large chunks of time becoming acquainted with interactive video (IAV). While much more is left for me to learn before acquaintanceship becomes mastery, the endeavor already has influenced my thinking. The experience has been sufficient to cause me to form opinions and arrive at tentative conclusions about the IAV phenomenon. I say "tentative" because the votes on IAV are not all in and future developments can only be projected—not assured. Anyway, I couldn't help but notice that my IAV learning experience, however incomplete, has generated opinions which don't seem to be held by all and thus may be worth sharing. Be forewarned, not all of my insights are rosy. For example, this instructional designer believes that . . .

- The potential of IAV in education is being oversold. IAV is not and probably never will be an "all purpose" instructional delivery system.
- IAV is becoming a buzz word, and being involved with IAV is the Faddish thing to do.

Roberts A. Braden is Assistant Director of Learning Resources and Head of Instructional Development at Virginia Polytechnic Institute and State University, Blacksburg, Virginia.

- IAV is too costly. Unless we are very lucky, IAV will have lost its allure before the public school systems can afford it.
- The general fascination over the technological elegance of IAV is obscuring the real problems that we educational innovators face in bringing this very promising development into the educational mainstream.

What Are We Doing?

Let me develop the last idea first. Ask yourself, what are we doing as we try to communicate with each other about this new instructional resource? Take the professional literature as an index. Are you, like me, seeing mostly articles that are about hardware systems? For instance, typical IAV authors don't choose a topic that would contrast different branching schemes; they are more apt to pick a subject that differentiates between videotape and videodisc systems. We tend to talk and write about Level II and Level III performance as a function of the machines, not as characteristics of information delivery.

When an IAV project is described we usually discuss whether disc control was created with an authoring language, an authoring system, a standard computer language, or some combination. But we don't often describe the kinds of author messages that we

use in student feedback or the level and quality of answer judging. In short, the communication that is taking place seems to be more concerned with the hardware technology than with the instructional and instructional design aspects of IAV. That is, of course, only my opinion. And, thank goodness, as with most generalizations, there are exceptions.

Unasked Questions

At a recent professional conference (NSPI) I received many questions about the equipment and the authoring tools we're using in our instructional design work. Nobody asked about our design process. Nobody asked how we determine student entry level. Nobody asked how we measure student progress. Nobody asked how we decide upon the amount and nature of student-system interaction. Nobody asked how appropriate our subject matter proved to be for delivery via IAV. Nobody asked how and why we designed our screen displays. Indeed, nobody asked about many other things that really ought to be of primary importance to any instructional technologist, and certainly ought to be the current shoptalk topics for instructional designers.

What Happened to ITV?

Turning back to my earlier statements that IAV is being oversold, is Faddish, and currently unaffordable in schools, I am reminded of the days when the A-V experts were hyping instructional television (ITV) as the educational panacea. The bandwagon was full of optimists. "Instant visual instruction," they proclaimed, "available at the flick of a switch." The mediated classroom was assumed by those experts to be just around the corner, and they also assumed a flood of quality educational TV programming to be imminently available in every classroom. Wrong. We all know that things didn't work out to fulfill those early expectations. We also can identify when and how things went astray. Consequently, maybe we should try to draw a couple of lessons from that chapter of recent history.

First, the instructional design technology lagged too far behind the electronic technology. When the glamour wore off, there wasn't enough good instructional product to maintain interest. Part of my message is that IAV hardware technology is ahead of the design technology to support it.

Second, ITV was expensive. The Ford Foundation primed the pump with hundreds of millions of dollars, but neither they nor the mighty federal government were willing to quickly come up with enough money to equip every classroom in our thousands of public school systems. A third of a century later we don't yet have TV sets in all of this nation's elementary, secondary, or college classrooms—partly because we still don't think we can afford them. So? Well, remembering that cost was a critical factor for TV, consider that the technology for IAV is more costly yet. The "video" in IAV subsumes all TV expenditures, while "interactive" adds computer costs as well. Whereas one or two TV sets can serve a classroom of 30 students for group viewing, a pair of IAV learning stations would be not nearly enough for 30 students working individually. The logic of these facts seems to ask, "If we couldn't afford milk, how can we afford the cow?" But it isn't that simple.

Cost-Effectiveness

The educational system is unable to afford any expensive alternative *that does not give a fair return on investment.* Continuing the cow metaphor, we don't need to buy any *sour* milk. That isn't pessimism, that's pragmatism. Optimistically, we *can* afford IAV, if (and only if) we are able to capitalize upon its potential. To clarify: IAV can succeed only if we design programs that are cost-effective. Yes, we can have our cow but not unless she gives enough milk and thick sweet cream to pay for herself.

To seek cost-effectiveness, we designers of IAV software would do well to set for ourselves two management guidelines: (1) Restrict the use of IAV to those applications where this new medium has *instructional advantages,* and (2) Produce *educationally efficient courseware* or none at all. The

litmus test must be that either the teacher's job or student's task (or both) shall be made easier by IAV—otherwise, we cannot justify the price. I suspect that to achieve educational efficiency means IAV must be designed by people who are knowledgeable about computers, instructional television, learning theory, instructional design, interactive strategies, and the subject being taught.

Interactive video isn't going to make instructional design easier. As a consequence of IAV, learning may get easier. I hope so. However, preparing the learning materials can only get more difficult as the complexity of the learning system increases. And, believe me, properly designed IAV is the epitome of complex learning systems.

A New Species

The instructional design generalist of the near future will need to be more of a renaissance man or woman than ever before. Moreover, we can look forward to a new species of ID specialist—the IAV programmer. My guess is that there will be lots of IAV specialists within two or three years—a few will even reside in public schools and on college campuses. The question is, will they all be in government and industry ten years from now, or will public education also find a lasting place for this technology and its technologists?

Coming Up . . .

This topic cannot be covered adequately in a single column. In Part Two we will take a closer look at the instructional advantages of IAV and how the competent designer can capitalize on them. Until then . . . □

Educational Technology Columnists

Instructional Design

Interactive Video: A Formative Evaluation (Part Two)

Roberts A. Braden

The title of this series of columns suggests both a topic, interactive video (IAV), and a process, formative evaluation. The fact that the process of formative evaluation can be applied to the topic of interactive video (IAV) tells us that IAV is still in a formative, evolving state—a state where we and others may still be able to influence its development. Surely formative evaluation is a tool to be used toward that end.

In a narrow sense, to evaluate is to gather enough information about something so as to be able to judge its worth. Formative evaluation, however, offers the special opportunity for applying our considered judgments so as to redirect the way things are going— or at least of affirming that the course being taken leads in the right direction. Unless all of us in the field of educational technology continue to observe and to assess, and to analyze, and thus continuously evaluate the status quo of IAV, we are apt to let another promising instructional innovation get away from us.

In Part One of this mini-series, two guidelines for IAV course designers

Roberts A. Braden is Head, Center for Educational Media and Technology, East Texas State University, Commerce, Texas.

were strongly advocated. One was to produce only courseware that is *educationally efficient*. The rationale given was that of accountability by substitution, i.e., the increases in cost of instruction are seen as a trade-off for savings in time or effort. The second guideline was to use IAV only when it offers clear *instructional advantages*. No rationale was offered, and none is needed, other than to say it is a common-sense solution. IAV is expensive. Unless there is an identifiable, clear advantage over less costly delivery systems, why spend the extra money?

Instructional Advantage

Notice that both guidelines, if followed, become selection criteria. Selection or rejection of IAV as the instructional delivery system of choice is, dear friends, the first pressure point where instructional designers can influence the fate of this new technology. Doesn't it make sense, then, that we should be as concerned, as careful, as rigorous, as systematic, as efficient, as knowledgeable, as professional, as generally competent when we select or reject the IAV delivery mode as we are when we design IAV courseware? That is why it also makes sense to select IAV based upon one or more identified instructional advantages and the prospects of improving educational efficiency.

The term "instructional advantage" as used here refers either to a unique capability of IAV as an instructional delivery system or to a superior teaching/learning result which can be obtained by using that system. The word instructional is critical to the term. Other authors have not made that distinction, and numerous lists of simply "advantages of IAV" exist. The lists I've seen contain few purely *instructional* advantages, but mostly they are an enumeration of the physical characteristics and capabilities of the equipment.

It is nice for an instructional designer to know that a particular IAV configuration "can accept various input systems," or "can respond in a fraction of a second" (stated in microseconds, naturally). Nice, but not essential to instructional design. Certainly not in a general sense. Actually, response time *is* a design factor, and we end up designing instruction to fit the machine at hand and whatever we might plug into it. These and a host of other design-to-survive factors will be set aside for a while longer. For now it is enough to keep instructional and equipment concerns as separate as possible so that we may have a clearer view of the kind of IAV we would really like to see adopted.

IAV and CAI

Before itemizing the best features of IAV, let me point out that the most important instructional advantages of IAV are all related to characteristics of computer-assisted instruction (CAI). Then, at the risk of leaving out a characteristic which is everyone else's favorite, here is my six-pack of IAV/ CAI instructional advantages. The first three are learner-oriented. The last three are teacher-oriented.

1. Pacing falls naturally under student control.
2. Immediate feedback is easy to provide.
3. Branching options facilitate meaningful individualization.
4. The system can be both more

patient and more persistent than a human teacher.

5. Recordkeeping and data-gathering can proceed automatically and concurrently with learning.
6. The instruction can be made available on short notice with little inconvenience to the teacher.

Why bother with IAV if CAI has the same advantages? As its name suggests, interactive video departs from other interactive CAI options *at the point where video images are added*. Excuse me for going back to basics, but some things are easier shown than described. In selecting IAV projects, then, we are on the lookout for an instructional need that calls for CAI plus "pictures." If there isn't a *real* need for pictures, CAI is sufficient.

The IAV selection process should assure that the project will prosper by incorporating features from the above-listed CAI advantages. To turn them into IAV/CAI advantages, the anticipated instructional design should capitalize upon the plain and fancy imagery which is possible with today's video technology. A designer can have stills, motion (slow, fast, normal, and reverse), animation, overlays, zooms, split screens, and every other kind of special view or image manipulation that might strike the imagination of the local special effects expert. If an example can be conceptualized, it probably can be visualized.

With IAV there are so many visualizing options that there is a strong temptation to get fancy for the sake of strutting our stuff. Keep in mind that what IAV does best is supplement CAI with (a) simple options for showing *motion* image sequences, or (b) access to a massive collection of still pictures.

Systematic Selection

The old pros recommend that instructional media be selected systematically. The typical selection system involves following a sequence of decision steps and calls upon information about the learning environment, equipment characteristics, learner attributes, and so forth. Someday we may have a computerized expert system to aid us in matching the variables to obtain the best fit. Until then, my suggestion is that each designer choose any sound selection procedure with which he or she is comfortable. To that process the designer ought to add appropriate selection tests when IAV is under consideration. To my way of thinking, at least three tests are needed.

My first IAV selection test is the image test. Ask: Will the instruction be more powerful, more efficient, or otherwise better if motion is shown or a large collection of images is accessible? If not, don't go the IAV route; if so, proceed.

The second test is the CAI test. Ask: Will some or all of the six key advantages of IAV/CAI be critical goals for the project? This is the toughest test and requires the most objectivity. As part of this test you may also have to ask: Will we be able to convince the individuals who benefit from IAV that the advantages apply to them? That means don't invent a superior product for which there are no consumers. Remember the Edsel? The lesson of the Edsel was that a wonderfully designed machine could be built that nobody wanted to buy. The designers drew ridicule rather than praise, not for poor design, but for being out of touch with the customers.

The third test is the impact test. Ask: Will the potential project fully exploit the power of your equipment, the talent of your design and production staff, and the higher order learning skills of the students? Unless the answer is affirmative or partly so, look for an easier or cheaper solution. Don't spend gobs of time and buckets of money designing IAV composed of trivial memory drills, lessons that teach topics out of the curricular mainstream, and simple subjects that can be taught as quickly and easily by another way.

That's it. As easy as one, two, three. One, the image test. Two, the CAI test. Three, the impact test. If your project needs images, will benefit from the power of CAI, and will have impact on the overall instructional program, you have a likely project for IAV. □

A Videodisc Approach to Instructional Productivity [1,2]

**Larry Peterson, Alan M. Hofmeister,
and Margaret Lubke**

There is no question that technology provides the educator with alternatives. Also, there is little doubt that certain technology-based applications, such as word processing and computer programming, enhance the societal relevance of curricula. The knowledge explosion and the associated growth of new technologies required curricular changes and promised tools to increase instructional productivity. This latter issue of using technology-based tools to increase instructional productivity is the topic addressed in this article.

Even a cursory review of research fundings of the past 20 years reveals considerable disagreement on the productivity benefits of different technological tools. For example, Baker (1978) in discussing one of the oldest tools, computer managed instruction, stated, "...the teacher is freed of many clerical-level tasks and is able to devote more time to instructionally related tasks. The greater the degree of computer involvement, the greater the gain in usable time" (p. 385).

On the other hand, using an econometric approach to assess productivity, Levin and Meister (1985) made the following observations regarding instructional technology in general and computer-assisted instruction in particular: "...the generic failure of educational technologies has been due largely to a misplaced obsession with hardware and neglect of the software, other resources, and instructional settings that are necessary to successful implementation" (p. 9).

Levin and Meister (1985) did, however, report that productive programs were "technically feasible," as evidenced in such products as Sesame Street, the NOVA series, and several Plato CAI programs.

Larry Peterson is Director, Vocational Education, Logan City School District, Logan, Utah. **Alan M. Hofmeister** is Director, Technology Division, Developmental Center for Handicapped Persons, Utah State University, Logan, Utah. **Margaret Lubke** is Research Associate with the same division.

Marche (1987), in one sampling of opinions of administrators concluded: "These school administrators had serious doubts about how much computing technologies are going to contribute to teaching effectiveness, with principals having a higher level of reservation" (p. 29).

There is evidence to suggest that technology has the potential to increase instructional productivity. But evidence also suggests that we have considerable work to do in identifying the most effective ways to design and implement technology-based programs that ameliorate rather than compound the problems of instruction. The following discussion describes a technology-based program and its role in enhancing instructional productivity in a demanding instructional setting.

Videodiscs and Schools

The videodiscs selected for this project were the Core Concepts Mathematics Program (Systems Impact, Inc., 1986) which teaches fractions, decimals, and word problems. Table 1 lists the major concepts covered by these videodisc programs. These programs resulted from a major cooperative effort by private industry, universities, and several school districts. Some videodisc-based programs are designed for "Level 3," that is, the videodisc player is driven by an interfaced microcomputer. The Core Concepts programs are validated for "Level 1" use; that is, the videodisc player is used without an external microcomputer and is driven by remote control. In group instruction the remote control is operated by the teacher. For individual learning-station use, the remote control is operated by the student.

In reviewing the Core Concepts programs, several appealing attributes were identified. One of the purposes of the project was the verification of these attributes. The attributes were as follows:

1. The programs could be used in both group and individual instructional settings and were compatible with the existing practices of most effective teachers who teach in both settings (Rosenshine and Stevens, 1986).
2. The programs emphasized courseware and de-emphasized hardware. Instead of one system per student or one per teacher, it appeared to be practical to use one system per building in most cases.
3. The programs required modest investments in staff development to effectively implement. Two hours of training were suggested. Much of the training had little to do with the technology, but focused on procedures to ensure that teachers effectively spent the increased time available in contact with students.

4. The initial review of implementation and maintenance costs suggested the programs would be very competitive. The additive costs were less than ten cents per student hour. Some of the alternate technology-based programs were costing close to a dollar per student hour for hardware, courseware, and maintenance.

5. The programs had an excellent history of internal and external evaluations, field tests, and controlled experimental findings showing that the products delivered substantive, positive changes in student achievement and attitudes (Hasselbring, Sherwood, and Bransford, 1986; Carnine, Engelmann, Hofmeister, and Kelly, 1987; Meskill, 1987).

6. The programs were clearly designed to enhance and complement the teacher's strengths. It appeared that teachers who had been spending large amounts of instructional time at the chalkboard could easily double the amount of time they had to spend on the classroom floor monitoring, praising, helping, and supporting students. With the teacher controlling the videodisc player with the remote control from any place in the classroom, there was no confusion about the controlling role of the teacher or any threat of technology replacing the teacher.

7. The video output was "broadcast" quality and had the potential to dynamically and professionally present concepts to a demanding school-age population used to commercial television.

The Productivity Field Test

The videodisc-based programs were used in all of the 11 fifth-grade classrooms in Logan, Utah. Utah school districts are under considerable pressure. For example, the state has one of the lowest per pupil expenditures in the U.S. Also, it has a rapidly expanding school age and aging population, which limit the tax base available to schools. Previously, the state economy depended heavily on its energy, steel, and agricultural industries, all of which have not fared well in recent years. To compound this, Utah is a state with high expectations for its school system. That is, the state has the highest proportion of its high school graduates going on to higher education, and its standardized test results exceed what one would expect of an educational system funded so modestly.

The state has recognized the need to prepare school graduates for employment in a competitive information age, and was the first state in the nation to mandate computer literacy requirements for all applicants for teaching certificates. Thus, the state is also searching for ways to use technology to increase instructional productivity. This project was one of several efforts to evaluate the role of computer, video, and telecommunications technologies in instructional productivity.

The average number of pupils in each of the 11 field-test classrooms was 33. The average for the previous year was 31.5. Of the 33, an average of four in each class were Chapter I and special education pupils who participated in all the videodisc-based instruction and whose test results are included in Tables 2 and 3. Clearly, the instructors were under some pressure with these large and diverse classes. This pressure had been increasing steadily for several years. In earlier years, class sizes were in the 20-25 range, and most exceptional children had been taught in segregated programs. The question of concern was: Can the videodisc-based programs help the teacher achieve high levels of student mastery for these large and diverse classes?

Implementing the Programs

The videodisc-based programs were implemented in January 1987, after a two-hour training session. The 35-lesson fractions program and the 15-lesson decimals and percents program were used in all 11 classrooms. The word problems program was an optional program, and only implemented by some of the teachers after the standardized testing was done in early April. The word problem results are not reported or reflected in the standardized test scores. It was the first time the word problems program had been used, and there was some concern that it might be more applicable at a higher grade level. The teachers who did use the program experienced success and plan to continue its use.

All the Core Concepts programs reserve every fifth lesson for diagnostic mastery testing and review. If these fifth-lesson mastery tests indicate problems, then reteaching is conducted before the group moves on. The fifth-lesson mastery test results therefore underestimate the levels of mastery (see Tables 2 and 3). To facilitate program evaluation, additional comprehensive, criterion-referenced pre- and posttests were given for both the fractions and decimals programs. A few weeks after the videodisc programs had been taught, the district's annual standardized testing program was conducted.

Table 2 shows the mastery testing results for the fractions programs for each of the 11 classrooms. The pretest results clearly indicate a population with minimal skills. The fifth-lesson mastery testing and the posttest results show a consistent pattern of success across all classrooms. Table 3

Interactive Video

Table 1

Major Concepts Covered by Systems Impact Programs

MASTERING FRACTIONS

- Pictures of fractions
- Fractions more than, equal to or less than 1
- Equivalence
- Fractions on a number line
- Common numbers
- Adding and subtraction fractions
- Simplifying fractions
- Mixed numbers
- Dividing by fractions
- Discrimination - add, subtract, mutiply and divide

MASTERING DECIMALS AND PERCENTS

- Reading and writing decimals with tenths, hundredths, and thousandths
- Adding and subtracting decimal numbers
- Multiplying decimal numbers
- Dividing to change fractions and mixed numbers into decimals
- Dividing decimal numbers
- Converting fractions, decimals, percents and whole numbers
- Rounding decimal numbers
- Solving simple percent word problems

MASTERING RATIOS (Word Problems)

Equation Skills
- In-place multiplication
- Solving equations
- Estimating ratio numbers
- Rewriting equations
- Writing equations from tables

Word problem Skills

- Sentence analysis
- Writing equations from word problems
- Verbal restatement of ratios
- Linguistic variations
- Percent problems with embedded 100%
- Rate problems
- Averages
- Unit-conversion problems
- Table problems
- Increase problems
- Two-question problems

Table 2

Average Percentages for Fractions Tests

Class	Pretest	Mastery Tests							Posttest
		5	10	15	20	25	30	35	
1	17.4%	96.8%	95.9%	90.7%	96.6%	91.6%	94.3%	74.3%	90.1%
2	16.8%	95.4%	95.5%	89.9%	95.6%	91.3%	91.2%	78.1%	87.1%
3	16.8%	97.4%	93.0%	92.7%	95.6%	89.9%	96.1%	77.1%	92.2%
4	9.8%	88.5%	92.0%	89.7%	87.2%	92.2%	85.7%	71.2%	78.1%
5	12.6%	96.2%	94.2%	90.1%	89.8%	94.2%	92.7%	75.6%	87.8%
6	8.5%	97.3%	95.6%	87.1%	93.0%	91.9%	90.3%	82.1%	86.0%
7	11.7%	98.5%	96.0%	93.0%	96.6%	96.0%	94.4%	82.1%	88.3%
8	13.1%	99.4%	94.6%	90.0%	96.2%	85.6%	93.7%	86.5%	88.9%
9	7.9%	95.9%	92.6%	88.9%	87.2%	83.4%	74.5%	58.4%	96.1%
10	14.9%	94.9%	89.6%	82.7%	85.7%	80.0%	84.4%	54.3%	81.5%
11	12.2%	96.8%	93.2%	88.5%	89.9%	84.3%	79.5%	64.4%	84.4%
Summary	13.1%	96.1%	93.8%	89.1%	92.2%	89.0%	88.8%	73.0%	87.3%

Note. This table lists the results of the comprehensive pre and post tests and the imbedded mastery tests for each of the 11 classes.

shows the same test information for the decimals and percents program; again, the same consistent pattern of skill acquisition is in evidence. A review of the curriculum covered by the two programs listed in Table 1 reveals that many school curricula do not require mastery of several of these concepts at the fifth grade level. While many of these skills are introduced at the fifth grade level, mastery may not be required until the seventh or eighth grade.

Each year these teachers have experienced larger and more diverse classes, and the field-test group was no exception. In an effort to gain extra evaluation data on the effectiveness of the classrooms, the standardized test data from the fifth grades of the previous year were compared with the data from the field-test population. One major problem was encountered in making this comparison. Some of the advanced concepts taught to the field-test group had not been taught to previous students and were not included in the form of the standardized test used. In order to have the standardized test better reflect the curriculum, we would have had to use a form designed for the junior high school rather than the elementary

school. This problem would suggest that any differences between years would be conservative measures of differences if all other factors were equal.

In making comparisons with the previous fifth grades taught by these teachers, we reviewed the sub-skill mastery data supplied on each of the major topics in the computerized test data (CTBS Math test form U). The field-test treatment covered approximately one quarter of the year's instruction and the five topics shown in Table 4. In reviewing the other topics, such as measurement, geometry, and number theory, there appeared to be only small differences between the two years. These small differences did, however, favor the previous year, not the field-test population.

A review of the standardized test data in Table 4 indicates a number of consistent trends. First, the percentage of students classified as mastery or partial mastery appear to be consistent with the criterion-referenced data shown in Tables 2 and 3. Secondly, the trend on each topic clearly and consistently favors the field-test group. Third, despite the fact that the field-test group covered more

Table 3

Average Percentages for Decimals Tests

| Class | Pretest | Mastery Tests | | | Posttest |
		5	10	15	
1	6.0%	90.2%	74.5%	73.7%	81.5%
2	5.6%	88.9%	82.1%	84.8%	82.8%
3	6.4%	84.6%	74.3%	77.4%	87.0%
4	8.5%	84.9%	77.5%	89.6%	77.2%
5	5.3%	88.5%	86.3%	83.3%	84.7%
6	5.5%	89.8%	79.6%	89.3%	83.9%
7	6.8%	89.3%'	77.7%	86.5%	82.6%
8	3.4%	76.1%	63.9%	89.3%	67.1%
9	4.1%	78.0%	74.6%	83.3%	72.5%
10	6.7%	79.0%	74.5%	*	*
11	*	82.2%	72.6%	78.7%	80.1%
Summary	5.6%	85.4%	75.1%	83.9%	79.6%

Note. This table lists the results of the comprehensive pre and imbedded mastery tests for each of the 11 classes.
* Missing data.

content, was a little larger and a little more diverse in that there were more special education pupils included, the percentage of students failing to achieve full or partial mastery dropped from an average of 13.4 percent to 8.4 percent. The reduction in the failure rate is even more remarkable when one considers that the whole school district experienced highly abnormal levels of absenteeism during the field-testing period because of a particularly bad epidemic of influenza.

Teacher Reactions

After using the fractions and decimals programs, all participating teachers completed a questionnaire.

The questionnaire required the teachers to identify the strengths and weaknesses of the programs.

The hardware. No negative comments were registered on the hardware. All teachers classified the videodisc player as "easy to use" on a four-point scale from "easy to use" to "hard to use."

Level of teacher support. The most common comments related to the reduction in pressure through reduced time in preparation and increased help in the instructional presentations. These comments, in essence, supported the notion that the technology was giving them more time for increased instructional contact with students. Typical comments included:

Table 4

CTBS Math Subscores

CTBS MATH SUBTESTS	1985-1986			1986-1987		
	Mastery	Partial Mastery	Non Mastery	Mastery	Partial Mastery	Non Mastery
MATH COMPUTATIONS						
Addition of Decimals and Fractions	50%	24%	26%	60%	26%	14%
Subtractions of Decimals and Fractions	62%	26%	12%	71%	21%	08%
Multiplication of Whole Numbers	77%	16%	07%	84%	12%	04%
Multiplication of Decimals and Fractions	58%	33%	09%	69%	24%	07%
Division of Whole Numbers	63%	24%	13%	70%	20%	09%
Summary	62%	25%	13%	71%	21%	08%

Note. This table lists percentages of 1985-1986 fifth grade students and 1986-1987 fifth grade students who demonstrated mastery, partial mastery, nonmastery on fraction and decimals computation skills. The information was taken from CTBS math subtest scores.

1. "Allowed for more interaction with students."
2. "Much of the preparation and presentation taken care of by the program."
3. "Allowed me to develop other roles such as monitoring students."
4. "Gave me time to work with more students individually."

Level of student support. Teacher comments on students' reactions were of two types: Some comments addressed the instructional design; others addressed the motivational nature of the delivery. Comments included:

1. "Helped students because of pacing, reinforcement of skills."
2. "Helped with presentations when kids were absent." (This related to the use of the system as an individual workstation for students who had missed lessons.)
3. "Maintained student attention because highly motivating; pacing; TV; they love the 'shiny' disc."
4. "Kept student attention because of the interactive features."
5. "Helpful because of the ability to remediate (catch up) students with excessive absences."
6. "Helpful to the students because the presentation of the math was clearer and more graphic."
7. "Helped students because of the teaching approach—a little at a time, step-by-step, guided practice and questioning."

Changes in the classroom atmosphere. Teachers

were asked to comment on any observed positive or negative changes in the classroom atmosphere. No negative comments were registered. Examples of the comments included:

1. "Business-like attention to work really strengthened."
2. When kids listen to program, they know they have to stay on task." (This "on-task" comment was a common theme in comments and is probably due to the combination of an engaging presentation and the fact that the teacher is now spending far more time on the classroom floor monitoring and helping.)
3. "The atmosphere changed; math became more enjoyable for everyone."
4. "The atmosphere became more relaxed and less frustrating."
5. "The kids were more excited."

Future plans for the videodisc programs. Teachers were asked to comment on their future plans. All teachers indicated that they would continue to use the program.

Conclusions

One stereotypic line of thought on the use of technology is that we use it because we must create an information-age environment in the classroom. For this reason we have to tolerate complex and expensive devices, a total restructuring of the classroom, and a depersonalization of the instructional environment. The findings of this project serve to contradict this stereotype.

Both the student achievement data and the observation of teacher and students' attitudes would suggest that:

1. The videodisc-based programs increased the instructional productivity of the teachers in that they were able to teach more content to mastery and meet the needs of large and diverse classes with less pressure.
2. The technology did not require a massive reorganization of the classroom because the selected videodisc-based programs support-

ed both group and individual instructional settings and were implemented with comparatively modest staff development and fiscal costs.

3. The technology helped the teachers increase the level of personalization of the instruction in that teachers were more knowledgeable of individual performance and were spending far more time in contact with individual students. □

References

Baker, F.B. *Computer Managed Instruction.* Englewood Cliffs, NJ: Educational Technology Publications, 1978.

Carnine, D., Engelmann, S., Hofmeister, A., and Kelly, B. Videodisc Instruction in Fractions, *Focus on Learning Problems in Mathematics,* 1987, *9*(1), 31-52.

Hasselbring, T., Sherwood, B., and Bransford, J. *An Evaluation of the Mastering Fractions Level-One Instructional Videodisc Program.* Nashville, TN: George Peabody College of Vanderbilt University, The Learning Technology Center, 1986.

Levin, H.M., and Meister, G.R. *Educational Technology and Computers: Promises, Promises, Always Promises.* Stanford, CA: Stanford University, Institute for Research on Educational Finance and Governance, 1985.

Marche, M.M. Information Technologies in Education: The Perceptions of School Principals and Senior Administrators. *Educational Technology,* 1987, *22*(4), 28-31.

Meskill, C. Educational Technology Product Review: Mastering Ratios. *Educational Technology,* 1987, *27*(7), 41-42.

Rosenshine, B., and Stevens, R. Teaching Functions. In Merlin C. Wittrock (Ed.), *AERA Handbook of Research on Teaching (4th Edition)* (pp. 376-391). New York: Macmillan Publishing Company, 1986.

Systems Impact Inc. *Core Concepts in Science and Mathematics.* Washington, DC: Systems Impact Inc., 1986.

Notes

1. This project was supported by funds from the Utah State Department of Education.
2. Thanks are expressed to all the fifth grade teachers and their principals who participated in this project. The help of Gary Carlston, Curriculum Director, Logan City School District, is gratefully acknowledged for his support in planning and evaluating this project.

Part Five

Emerging Interactive Technologies

History in the Making: A Report from Microsoft's First International Conference on CD ROM

David Rosen

Microsoft's First International Conference on CD ROM (there will be a second, next year in San Francisco) made history in two different regards. One was planned in advance by the organizers; the other sensed, but not at all fully anticipated, either by Microsoft or most conference participants.

The first historic aspect was the gathering, one thousand strong, of the new media tribes—software developers, consumer electronics manufacturers, book publishers, database producers and electronic information services, telcos, computer makers, interactive videodisc pioneers—united by a common interest in CD ROM, and inspired (like a previous congregation on Max Yasgur's farm) by the impressiveness of its own assembly.

The second came in a plenary session the morning of the second day, when participants witnessed the announcement, by David C. Geest (pronounced "haste"), chairman of Philips' Home Interactive Services Group, of what may arguably constitute a new industry—a medium called CD-I.

Even without the dramatic announcement by Philips, the conference would have produced enough important developments to spur CD ROM's momentum as an emergent commercial phenomenon. Among them:

- Sony's U.S. CD manufacturing subsidiary, Digital Audio Disc Corporation of Terre Haute, Indiana, and Digital Research founder Gary Kildall's Knowledgeset Corporation (formerly Activenture) of Monterey, California announced an agreement to form a joint venture to provide "one-stop" serfice for formatting, mastering, and replicating CD ROM software.
- Dialog Information Services stated that "this year we expect to announce our first CD ROM products."
- Microsoft chairman Bill Gates made clear that his company will make a major commitment to CD ROM, with a surprisingly strong emphasis on consumer applications.
- Dun & Bradstreet demonstrated a CD ROM version of its *Million Dollar Directory*, with retrieval software developed by LaserData.
- Discovery Systems of Columbus, Ohio, headed by CompuServe founder Jeffrey Wilkins, announced it will manufacture CDs, CD ROMs, and 12-inch videodiscs in production quantities in 1986; it also announced its agreement with Battelle's Software Products Center to jointly develop compact disc publishing services.

Most significantly, however, and somewhat to the chagrin of limelight-seekers developing professional database applications, the conference was picked up and shaken by the revelation of CD-I and later by two presentations by Stan Cornyn of The Record Group—a former Warner Communications executive who strongly influenced the concept and development of CD-I as a world standard for interactive multimedia programming.

What Is CD-I?

Compact Disc-Interactive, or CD-I, is a set of specifications, the final details of which are still being drawn up. Backed by Philips and Sony (the partners in the development of the compact audio disc and CD ROM standards) with other Japanese manufacturers (notably Matsushita) on board, CD-I expands upon CD ROM in two important ways. By defining how data for video, graphics, and sound are encoded on the disc, it standardizes a multimedia format; and by specifying a microprocessor family (Motorola 68000) and operating system (CD-RTOS, based on OS 9, made by Microware of Des Moines, Iowa), it enables real-time applications such as entertainment and education/training, and ensures that CD-I discs carrying audio, video, text, binary data, and applications programs will work on all CD-I drives from all manufacturers.

As a product, CD-I is intended to piggy-back on the success of CD audio by providing a CD player which can also function as a "viewer" for interactive programming. Although target retail price at introduction ("early next year") is $1000, eventually it is expected to be priced at approximately $200 above the price of compact audio disc-only players. Applications envisioned include:

David Rosen is Director, Consumer Electronic Media Program, LINK Resources Corporation, 215 Park Avenue South, New York, New York 10003. This article is based on a report written by the author for LINK clients earlier this spring; used with permission of LINK Resources Corporation, a subsidiary of International Data Corporation.

- **Entertainment**
  - "music plus" (music with text, notes, pictures, etc.)
  - action games
  - strategic games
  - adventure games
  - activity simulation
  - "edutainment"
- **Education and training**
  - do-it-yourself
  - home learning
  - interactive training
  - reference books
  - albums
  - talking books
- **Creative leisure**
  - drawing/painting
  - filming
  - composing
- **Work at home/while travelling**
  - document processing
  - information retrieval and analysis
- **In the car**
  - maps
  - navigation
  - tourist information
  - real-time animation
  - diagnostics

In addition, a CD-I player can also be used as an "intelligent" disc drive to another computer. Even though its imbedded processor is 68000-based, a CD-I player could be used, in theory (see below), to deliver text/data on CD ROM to an IBM PC workstation. The embedded intelligence may indeed prove useful in some pure-data applications (the frame buffering features of CD-I, for example, could assist much faster retrieval through creative management of indices).

It will also be possible to use recent and future audio compact disc players with a direct digital output to play CD-I discs, using an outboard decoding unit for replay of CD-I text and visual information.

CD-I players will contain their own control facilities, and will be able to use built-in keypads or keyboards, remote control units, and/or pointing devices. In most cases they will be connected to the user's existing audio and video equipment; or they can have their own audio and video facilities.

A Unique Medium

In contrast to CD ROM, with its emphasis on ASCII text, and storage and retrieval applications, CD-I is an audio/visual-driven, real-time medium. Its video capabilities (it is six times more efficient for video data storage than CD ROM, due to the inclusion of a special video processor) extend to high-resolution modes which will be available on digital televisions available within the next two years, as well as compatibility with current NTSC, PAL, and SECAM standards. But because of the data transfer limitations imposed by the compact dic's rotation speed, full motion video is not achieveable (at least, not as yet), a distinct drawback in many education, training, and home entertainment applications.

Therefore, the visual presentation of CD-I will rely on still and "step-motion" video (slide shows) complemented by high-resolution graphics, including real-time animation, and various text presentation options. Simultaneous to visual displays, CD-I programming will feature sound at specified levels of quality corresponding to CD audio, regular high fidelity, AM radio, and telephone. The lower the sound quality, obviously, the more can be stored on a disc (up to 16 hours of audio programming, complemented by images, can be provided at telephone quality).

OS 9, not a household word in operating systems, was chosen by Philips because it was "the best real-time operating system available" according to Richard Bruno, manager of Corporate Planning/Technology for Home Interactive Systems. Microware is a 10-year-old company with about 40 employees and 200 licensees for its real-time dedicated operating systems. They are imbedded in automatic teller machines, industrial automation systems, communications processors, and personal computers made by Tandy (the Color Computer), Fujitsu, and Hitachi.

Microware president Ken Kaplan told us the key attributes of OS 9 are its highly-portable Unix-type environment, ability to be placed on ROM, and very economical memory management characteristics. Speaking personally, he doubts that a rival Intel-based architecture for a CD-I viewer will emerge, because of the Intel family's bus limitations for handling audio, and the possibility that IBM is simply not as advanced in the technologies of multimedia encoding and decoding as Philips, Sony, and other consumer electronics giants, among other factors.

"Time Machine" and Other CD-I Programs

Stan Cornyn is president and owner of The Record Group, a production company assembled specifically for the new medium and backed by Philips subsidiary Polygram Records and "at this time"—to help get the medium started—by Philips Home Interactive Systems. In presentations at the Microsoft conference, supplemented by follow-up conversations, Cornyn lent imaginative substance to the kinds of programming made possible by CD-I. There are currently about 20 titles in some stage of planning or production at The Record Group, although "probably 10 will fall off the truck."

The furthest along, called "The Time Machine," provides a good example of the possibilities of "edutainment" on CD-I. Merely a history of civilization, the program might be described as a map-based multimedia history database. The user chooses a year, and a "map of the year" appears, along with headlines describing the current events and characteristics of the civilizations of the time. Want more information? Menus allow more detail on what and why; or click on a symbol and get a multimedia presentation on that subject. Viewed passively, the program cycles through the years at one per second, showing the ebb and flow of nations with voice-over "news commentary."

Another program, one-third completed, has the working title "Princess Di Is Related to Chiang Kai-shek." Based on a genealogical database of rulers and their dates, countries, predecessors, siblings, and spouses, it can be used to retrieve information, or played as a maze game. A third takes off from the classic Adventure game, but puts "real characters in James Bond-like adventures," adding "elements of literature" such as "characterization and suspense." Now in production is a tour of London, enabling the user to walk the streets, turning left or right with the corresponding audio-visual results—or entering a "fourth dimension" wherein lie the sights and sounds of London in Shakespeare's time, the Tudor period, or Celtic times.

Philips promises to provide the special "authoring systems" needed for post-production formatting of data for CD-I discs "by October 31st." The Record Group will be the first "designated recipient"—meanwhile it operates the only "designated CD-I pilot studio," with IBM PCs modified to run the OS 9 operating system. "If everything else is in place," says Cornyn, referring mainly to the authoring system availability, "we could have 'The Time Machine' ready by the first quarter of next year." He sees software distribution as "the damnedest thing we'll face for awhile."

Cornyn estimates the cost of producing CD-I programs as "at most $250,000 per title—using graduate student labor." The magnitude of a hit needed to break even (not counting the development cost of setting up the Group): "about 25,000 copies."

The Record Group's capacity to take on new productions during the next year is "comfortable up to about 50 programs where we're involved in the creative design part of it," says Cornyn, "more if we're just crunching someone else's data." The skill most in demand: "the ability to think fresh."

Compatibility with Current CD ROM Products

As followers of the commercialization efforts of the CD ROM industry are aware, most current implementations are compatible only with a given operating environment. Common database access elements, such as where and how files are identified and how files are opened to retrieve data from the disc, needed for interchangeability of a given disc title among different drive/computer operating system combinations, are not standardized by the "Yellow Book" specifications defining CD ROM.

Since this situation severely limits the opportunities for publishers to address broad end-user or institutional markets without separate mastering of alternate versions of each title, it has limited the willingness of information providers to create products on the CD ROM medium. Therefore, a group of vendors (composed of Reference Technology, Digital Equipment, TMS, LaserData, Xebec, Microsoft, Apple, Hitachi, Philips, 3M, VideoTools, and Yelick, Inc.) calling itself the "High Sierra Group" was formed to propose a *de facto* standard.

CD ROM discs can be made upward-compatible with CD-I, but only when the High Sierra group determines a *de facto* standard (which is accepted and implemented) containing a "bridge" to CD-I file access conventions. To ensure that this bridge is in place, High Sierra postponed its announcement of a proposed standard, most recently scheduled for March 31. It is most likely that such discs would contain data only, with CD-I specific retrieval packages provided separately on magnetic media.

The possibility remains that a rival standard based on the Intel processor family (the 80286 being the most likely), compatible on a standalone-basis with current and planned MS-DOS CD ROM applications, could emerge to challenge the Philips/Sony system—especially if IBM's presumed desire to dominate educational and home markets is manifested in a similar product to CD-I. Such a scenario is far from being the only challenge faced by CD-I system:

- Since no software/content for the system presently exists, the familiar chicken-and-egg dilemma is posed. Only The Record Group currently has programs in the works which are specifically intended for CD-I.
- Pricing of the system is going to be critical: the key will be to offer consumers the perception of an "enhanced CD record player." If the premium for CD-I's interactive capabilities is too great, and the available CD-I content too sparse at introduction, market acceptance could be difficult to obtain. (The big push is scheduled for the

Christmas season, 1987, with sample volumes available earlier in the year.)

Conclusion

What of the impact of CD-I on the **consumer market**? LINK believes that:

- CD-I market introduction efforts as a "home appliance" should overcome likely resistance in which entertainment has been traditionally viewed as either passive (e.g., radio/recorded music listening and television/video viewing) or [inter]active (e.g., video or computer games).
- Lack of recordability and erasability and absence of full-motion video will tend to limit CD-I from realizing its full market potential. (While videodisc picture quality is superior to VCR, the VCR revolutionized the consumer entertainment market due to its time-shifting capabilities and the availability of sufficient software.)
- On the other hand, CD-I will most certainly open doors to creative, more imaginative programming, leading to a new genre of entertainment and educational software drawing from the world of audio, video, and computers. This development in itself should provide strong impetus to product acceptance among early adopters.

The impact on the **overall CD ROM market**? LINK believes that:

- The prospect of a consumer electronic publishing medium with the same level of standardization that has enabled the Compact Audio system to achieve such rapid success can only facilitate the growth of an infrastructure supportive to the development of optical media publishing in general.
- Most current CD ROM applications, involving either text-only online databases migrating to locally-stored media or text-oriented internal data and relatively low-resolution images, will not be directly affected by the prospect of CD-I.
- Many relatively broad-market oriented professional and institutional publishers, who have been eyeing CD ROM as an entry into electronic publishing, will now opt to wait for CD-I and its greater multi-

media facilities and assurance of standardization.

- Educational and library buyers, for whom a single system to handle all electronic storage needs is highly desirable, and standardization crucial, will think twice before before making major commitments to CD ROM, while awaiting further word on CD-I.

LINK anticipates that CD-I will have a strong impact on the **training market** for the following reasons:

- The price of CD-I hardware will be substantially lower than interactive videodisc hardware at introduction (e.g., Sony View System at $7500 vs. CD-I at $1000) and will most likely decline rapidly during its first five years following introduction and increased penetration.
- CD-I is being initially targeted at the consumer market, and hence should be more user-friendly, ergonomically sound, and reliable than MS-DOS based interactive videodisc systems.
- Based on preliminary review of CD-I technical specifications, there appears to be a feeling among programmers that CD-I software could be easier to prototype than interactive videodiscs.
- For many in training, full-motion video may be an unnecessary luxury which can be appropriately replaced by CD-I stills, animations, or computer graphics, thus bringing down production costs.
- CD-I capacity using graphics and either telephone or AM audio recording will be far greater than interactive videodisc.

Conclusion

In sum, one cautionary note should be kept in mind: The anticipated high production costs that are likely to occur during the early phases of program production could limit the participation of many creative independent programmers. It should be remembered how, during the early years of the personal computer industry, independent software developers contributed to the introduction of a wide range and variety of applications. Such creativity should be nurtured for the long-term potential of the CD-I industry. □

Anticipating Compact Disc-Interactive (CD-I): Ten Guidelines for Prospective Authors

Michael DeBloois

Here it is less than a year since the Microsoft conference in Seattle, where the new Interactive Compact Disc Standard was announced (Rosen, 1986). No playback equipment is available; encoding technology is still in the embryonic stage; no practical software has been developed. Isn't it premature to give advice to prospective authors?

It's true, we know very little about CD-I and what it may become, but it is not too early to begin orientation and instruction of people who plan to author in this new medium.

Lowe (1981) evoked the familiar bromide of the blind men and the elephant to describe the multifaceted nature of the medium which will emerge from the CD-I standard. Lowe warned: "the more we concentrate on the similarities of CD-I to our favorite technology, the more we limit ourselves." Certainly it would do us little good to confine our thinking to the constraints of present and past technology. On the other hand, is it really possible to think about anything outside a known frame of reference? Concepts and ideas based in the present are the only intellectual tools we have for wrapping our minds around the future.

In this article I describe a number of different frames of reference through which we can look at the opportunity presented by CD technology. As a result of this approach, I will propose some general guidelines to help authors to begin thinking about a more creative role as a prospective author of CD-I materials.

As we move closer to the day when the CD-I standard is implemented, when playback equipment is available, and authors are producing CD-

Michael DeBloois is with Discovery Systems, 555 Metro Place North, Dublin, Ohio 43107. Formerly a member of the faculty of Utah State University, he edited *Videodisc/Microcomputer Courseware Design* (Educational Technology Publications, 1982).

based materials for a mass audience of consumers, we will, no doubt, see very specific authoring paradigms developed. There is little question but that these approaches will adopt many of the conventions presently being used by developers of interactive videodisc training materials. At an early stage of thinking about this new medium, however, I am of the opinion that we need to project beyond this and be widely divergent as we examine other authoring possibilities.

If my experience authoring videodisc-based programs has resulted in any authoring wisdom, it tells me we have much to learn about the medium before we can begin to proscribe authoring system specifics.

Guideline 1: *We should demonstrate the wisdom and maturity to first gain experience using the medium, evaluating what works and what doesn't, before attempting to define authoring protocols.*

Mark Heyer wrote an interesting chapter in *CD ROM: The New Papyrus* (1985) in which he describes a simple taxonomy of ways people gather information. He states there are three approaches: grazing, browsing, or hunting. As I clear my mind of images of gazelles, bears, and Hemingway in a red flannel shirt, the analogy begins to help me consider authoring strategies. Users seeking information on a compact disc may indeed wish to move intently and purposefully through all the content, as it is presented; chew on each part for a while and then move on to the next selection. Others may wish to jump around, picking choice morsels as they are spotted; then move on to the next morsel. There are also users who know, for the most part, what information they need, and want to get directly to what they are hunting.

The analogy falls short. There is more to an information gathering strategy than grazing, browsing, or hunting. For example, language learners want to *bathe* themselves in the culture and pick up subtle nuances as they are inundated in the sights, sounds, rhythm and knowledge of the language. Students of the sciences want to *discover* principles through an inductive process, and have the opportunity to come to their own conclusions. Hobbyists may wish to *gather* every relevant piece of information pertaining to their avocation down to the most detailed item; important information, unsubstantiated rumors, and irrelevant trivia make up the collection.

Guideline 2: *Authors must study the variety of ways users wish to receive information and learn* *to design presentation and interaction strategies that accommodate this diversity.*

David Hawthorne (1986) made an important observation at a recent conference that went something like this: "People who think that interactive video or CD-I have anything to do with American television viewing aren't paying attention. Watching television is a mostly passive activity which individuals engage in to escape."

Americans view the TV room in their home as a social decompression chamber where one can retrench, sit back, and adjust to degrees of lesser pressure. Sometimes a can of beer or a bag of corn chips helps ward off the bends.

Authors of interactive videodisc materials often thought they were creating a television-like medium. CD-I authors would do well to avoid this error. While we may not know what CD-I is quite yet, it is reasonably clear it will have little in common with TV viewing.

Guideline 3: *When authors think about designing and developing materials using compact disc technology, they should evoke and follow appropriate metaphors.*

Viewing the application of the technology through the frame of different metaphors helps us to think of different authoring strategies.

Let's try, for example, the "Electronic Coffee Table Picture Book Metaphor." Operating from this frame of reference, the author will attempt to develop materials for users who are seeking an aesthetic experience enhanced with a presentation of esoteric information. Only the highest quality of sound and image will be acceptable, and users will set the pace of interaction, ranging from lazy to leisurely. A very dense visual data base seems appropriate, supported by mood music, voice narration, and carefully written supporting text screens. Tempting as they might be, the wise author would be well advised to ignore using the 300,000 pages of text available.

Guideline 4: *Knowing when not to use the full range of CD-I capabilities for a given application may be as important as knowing what the capabilities are.*

Another metaphor expands the point. What strategies would be used to design and author "Antiques Across America; a source book to fine collections cataloged by type and period." A perfect CD-ROM application, some would claim. Nothing more than a data base requiring a relational search strategy. Well, maybe not. What a challenge for an author! How can the essence of a visit to a wonderful antique shop be recreated to be experienced at home? Vicarious browsing from room to room, picking up certain objects for closer inspection. Close-up shots of the patina of the wood, dust showing in the streams of sunlight slanting through window panes, the delicate color of oxidation on century old glassware. How can this be presented to create an electronic version of "going antiquing"?

Guideline 5: *Authors should avoid applying the same comfortable solution to each new project. An authoring strategy is a means for creating a desired end, not an end in itself. The development of different products will require the use of different authoring strategies.*

Consider the "Willie Nelson, a musical biography" metaphor. What kind of treatment would the author of a music-based CD-I product use? The music video format, which is already established, is rapidly loosing popularity and beginning to be viewed as a cliche. Anyway, who can top Michael Jackson's Thriller? What can be done with the life and times of a popular and colorful artist using the mix of media offered by CD-I technology?

Guideline 6: *Different media must be used as the dominant message carrying element of a media mix within a given program and across projects. For one project the video medium might dominate; for another, text might carry the message; for a third, audio could reign supreme.*

What authoring strategies would an author use to develop a "CD-I mystery home radio theater," or "the Illustrated Home Medical Advisor," or the "Worship at home, inspirational sermons with music and song" program? Each of these evokes authoring approaches and presentation strategies quite different from those floating around in the minds of traditional product developers.

One of the truisms that characterizes nearly every conversation I have had about CD-I is that the horizon is more vast than what we are accustomed to thinking about. Unless we really stretch to understand new and creative possibilities, we may fall victim to our intuitions and assume far too few capabilities.

Guideline 7: *Authors need to exercise thinking beyond present assumptions about media limitations. Our personal horizons need to be expanded to match those of the medium.*

Discovery Systems, my current employer, is a

market-driven company. Every major decision we have made since we have been in operation has been directed by our view of the marketplace. Having spent most of my career in an higher education environment, this has taken some serious personal adjustment. No longer is it enough to consider whether the product is well-designed, or whether it accomplishes what it set out to accomplish, or whether it is developed within the budget. CD-I will be market-driven, and authors and the companies for whom they work will live or die by how well the products they develop respond to market factors.

Guideline 8: *Authors must expand their view of what they develop to reflect an understanding of the impact of market forces.*

Companies and individuals who considered themselves to be designers and producers of interactive training or educational materials for videodisc-based programs have begun to realize that as the industry matures it is beginning to look more and more like a *publishing* industry—publishing in a form different from what we have known in the past, but publishing nevertheless. I believe that this will be the case for CD-I, but to an even greater degree. An industry will emerge where one segment provides information. Another segment will manipulate the information into an appropriate form, to be delivered to yet another segment which will encode the information onto the electronic media. Yet another segment will "print" or press the information onto optical discs to be distributed by another segment into the retail segment where the product will be sold. This is a very different view than that held by authors of interactive videodisc materials, who, for the most part, have adopted a *cottage industry mentality*.

Guideline 9: *Authors should begin to think of themselves working as a component of an electronic publishing industry, which will require carefully coordinated teams of specialists working collectively to produce and distribute their products.*

Authors who work with the complex media choices allowed by CD-I, within a competitive

and economy-minded electronic publishing industry, will be forced to adopt and use productivity tools. The sheer magnitude of controlling the vast amounts of information, and the requirement to interlace its thousands of elements in an aesthetic and creative way, and the need to work efficiently and hold costs down will require the use of much more sophisticated authoring approaches (expert systems technology) and computer-based tools than are available at the present time.

Guideline 10: *Authors should gain experience using the productivity tools presently available, and should begin providing input to appropriate sources to encourage the development of better and more useful tools for the future.*

Conclusion

Time passes very quickly. Seldom do people complain about having too much time for developing innovative approaches to meet the challenge of technological change. Now is the time for authors to discuss the challenges associated with authoring in a new medium. Compact Disc-Interactive is going to stretch our ability. We will need to deal with very complex issues in a creative and cost-efficient way; now is the time to think about and beyond the guidelines presented here. Those who propose to be authors should first become students. Success with this new medium will require new attitudes, a different set of assumptions, and innovative authoring skills. □

References

Lowe, L. CD-I Technology at the Focal Point: Part I and Part II. *Video Computing*, May/June 1986, July/August 1986.

Heyer, M. The Creative Challenge of CD ROM. In S. Lambert and S. Roplequet (Eds.), *CD ROM: The New Papyrus*. Redman, WA: Microsoft Press, 1986, p. 347.

Hawthorne, D. Personal luncheon conversation at Society of Applied Learning Technology Conference, Washington, D.C., September 1986.

Rosen, D. History in the Making: A Report from Microsoft's First International Conference on CD ROM. *Educational Technology*, July, 1986, *26*(7), 16-19.

CD ROM Joins the New Media Homesteaders

David C. Miller

Introduction

CD ROM stands for "Compact Disc Read-Only Memory." CD ROM is a remarkable new interactive publishing medium. CD ROM asserts, with Archimedes: "Give me where to stand, and I will move the earth."

"Wheres to stand," of course, are not GIVEN but must be EARNED. As a newcomer to the frontier, CD ROM has as yet no standing, no cleared ground of its own. CD ROM by necessity must take up the gruelling, hard-handed role of Homesteader.

CD ROM has its Band of True Believers, among whom I number myself. Since 1984, we have been digging up stumps and cutting brush in what we take to be CD ROM Territory. We have made our first plantings, and we have High Hopes. We have also bruises, calluses, and some hard-bought insights to share.

Educational technology stands fair to profit from CD ROM in at least two ways:

• In the first place, CD ROM is an important new tool for educational technologists (ET). Like any new tool, it must be mastered and its proper uses discovered. But CD ROM fits neatly alongside the microcomputer and the videodisc in the ET toolshed.

• In the second place, CD ROM is struggling to find and claim its rightful place in the world. From the outside, at least, it strikes me that ET remains engaged in that struggle, even after all these years. Perhaps we can learn something from each other.

I begin with a thumbnail sketch of the technology: how it works, and what it can do. There follows my hazy first notions of how CD ROM might be employed in ET. In conclusion, I outline the task-oriented, user-oriented, entrepreneurial perspective I believe is essential if CD ROM is to advance from Homesteader to Solid Citizen status.

David C. Miller is Managing Partner, DCM Associates, Post Drawer 605, Benicia, California 94510.

CD ROM: A Thumbnail Sketch

As the cliche goes, CD ROM is the entertainment industry's gift to the information industry. Philips-Sony introduced the Compact Music Disc in 1982. Earlier consumer music systems were based on analog recordings. CMD delivered noticeably superior sound quality based on digital recording methods.

CMD boomed, and soon became the most successful consumer electronic product ever marketed. A few million players and more than 70 million discs have been sold since 1982. Some observers believe CMD will crowd out conventional hi-fi records and tapes within the next few years.

Building on success, Philips-Sony introduced CD ROM in 1984. The technology is licensed worldwide to more than 100 companies, including giant corporations and many smaller and start-up enterprises.

CD ROM uses the same plastic blanks used for CMD, and can be stamped out in the same facilities used to mass-produce Compact Music Discs. CD ROM's physical recording specifications enable the storage of digital information files, computer programs, and images as well as audio recordings.

CD ROM players are more sophisticated versions of CMD players, and are designed to be used as peripheral drives with microcomputers.

Information is pre-recorded on a CD ROM master disc at the factory, from which unerasable replicas are mass-produced. Users can look at the information on a display screen or copy segments onto a magnetic disk for further processing. Nothing can be added to, removed from, or changed on the buyer's CD ROM replica disc. For that reason, it is called a read-only memory, just as a phonograph record is a read-only memory.

CD ROM Benefits

CD ROM offers a unique bundle of benefits. The disc is durable and long-lasting. It stores a great deal of information in a small volume. All information is stored in digital form, which means that disc contents are readily accessed and processed with a microcomputer. Finally, the discs can be replicated at a low unit cost, assuming that a significant number of copies can be sold.

CD ROM's durability stems from the disc's construction. Digitally encoded information is physically enscribed as sequences of pitted and non-pitted regions along a spiral track three miles long. A transparent cover seals the track from the surrounding environment.

Readout involves no physical contact with or wearing away of the information-bearing surface. The information surface is sprayed with reflective aluminum. The CD ROM drive readout head projects a laser beam through the transparent cover onto the information surface. Differential light reflection from pitted and unpitted regions is detected by the readout head.

The CD ROM disc measures 4.72 inches (120 centimeters) across and weighs 0.7 ounce. Yet this small disc stores up to 540 million characters (megabytes) of user information, plus 60 million characters worth of administrative and error-handling data. This translates into the equivalent of more than 200,000 pages of text or several thousand images or one to many hours of audio recordings, depending on the sound quality desired.

All information is recorded in digital form and can be read directly into a microcomputer's memory for display, copying onto magnetic disks, or further processing.

CD ROM data preparation costs are substantial, and vary depending on what must be done. Mastering a disc today may cost $4,000-$5,000. Discs can be replicated in quantity at unit costs of $60 to $5 each. These costs will decrease over the next few years, and compare quite favorably with print or magnetic disk costs when as few as 100 copies can be sold.

Readers interested in a detailed, comprehensive treatment of CD ROM should consult either of two works: *CD ROM: The New Papyrus* (Lambert and Ropiequet, editors, 621 pages, Microsoft Press, Redmond, WA 1986) or *Essential Guide to CD ROM* (Judith Paris Roth, editor, 189 pages, Meckler Publishing, Westport, CT 1986).

Using CD ROM in Educational Technology

No one knows how best educational technologists might employ CD ROM. Certainly, I claim no special knowledge. DCMA's emphasis is on CD ROM applications in libraries. The history of CD ROM in education has yet to be written.

I will mention one CD ROM educational product already on the market. Beyond that, I offer one or two sketchy notions of my own in an attempt to stimulate your imagination.

U.S., state, and local education agencies have developed large banks of standard, validated test items keyed to specific curricular objectives. Tescor, Inc. is compiling and republishing the best of these in the CD ROM-based "First National Item Bank and Test Development System." (Testcor, Inc., 461 Carlisle Dr, Herndon, VA 22070; 1-800-842-0077).

The National Item Bank contains several thousand criterion-referenced items and is updated every six months. Curriculum specialists and teachers can search and retrieve items by grade, domain, and keyword combinations. All items are identified by source, p-value, etc. Items retrieved can be edited before printing, and original local items can be inserted.

Once a test has been created, it can be printed out on a local printer or forwarded on floppy disk to Tescor for reproduction. Printed tests can be used with generic answer sheets and input to Tescor's Criterion-Referenced Test Scoring System. A full selection of analytical test reports can be produced on the local printer or forwarded and reproduced by Tescor.

The National Item Bank demonstrates that it is no idle fantasy to think of CD ROM as a new tool for educational technologists. How else might this powerful new medium be used?

Well, the U.S. Bureau of the Census has published its first prototype CD ROM discs. Beyond that, Census has announced it intends to use CD ROM as yet another publishing medium for the 1990 census. Suppose you could count on comprehensive Census data being available in every classroom, readily accessible via microcomputer?

What interactive instructional programs does that suggest? Comparative studies of localities, states, and regions? Scrutiny of family and household structures? Examinations of political participation? Remember: your students have full access to complete, current data. Interactive exploratory analysis of the "real world" can be substituted for anemic, constrained examples. What might be done?

Another hypothetical example: Here is a CD ROM disc which holds hundreds of audio files. Individual files are separately and rapidly accessible. Average playing times range between several seconds and one minute. Any and every musical instrument and combination of instruments can be represented. An array of tempos, chords, themes, etc., is on tap. Associated software enables students to edit and combine any two or more files, or to create their own files on magnetic disks. The sound quality available is limited only by local reproduction systems. How would YOU incorporate such a CD ROM product in your music education courseware design?

Or, how about combining text files from CD ROM with interactive images stored on videodiscs? What about interactive maps to teach geography? How about integrated text/audio files to teach foreign languages? How about using CD ROM to teach computer literacy or any of dozens of widely used microcomputer programs? Would an interactive disc with hundreds of college catalogs be useful to high school and undergraduate counselors? You take it from there.

CD ROM: The Homesteader Perspective

Congress passed The Homestead Act in 1862. The Act authorized the U.S. government to bestow 160 acres of federal land on citizens willing to clear and occupy new ground. The Act suggests a metaphor equally apt for CD ROM and, I suspect, for educational technology generally.

Earlier, I mentioned that CD ROM has its Band of True Believers, among whom I number myself. Too many of us, I fear, are blind zealots. We see and celebrate the many strengths and potentialities of our chosen medium. We too often fail to acknowledge the persuasive claims of older media that tilled the soil long before CD ROM came into the territory.

Our CD ROM zealots never acknowledge or too blithely dismiss the investments others made long since in things as they are. Sunken, irretrievable investments made in hardware, knowledge, experience, competence, and expertise. Investments made in media generally recognized as adequate, no matter what bright promises new contenders may hold out.

We are too eager to presume that any who do not instantly seize upon our offer to worship at our altar are inept, inadequate, inferior. We are too quick to believe that we should instantly be hailed as pre-eminent in whichever established forums we honor with our presence. We mount the cart without a horse when we prescribe without stopping to diagnose.

We do not realize that we must create our own opportunities, peddle our wares on their own merits. School budgets will not and should not automatically accommodate CD ROM. Perhaps what we offer is better suited to libraries, business organizations, community centers, union halls, or political action committees. If we bring new wine, we may be obliged to find or make our own bottles.

Who knows? Perhaps our proposed solutions are not worthwhile. Or perhaps we have yet to match our solutions with worthwhile problems and opportunities. We have thrust ourselves into the territory. It is up to us to homestead it, if we can.

Conclusion

I have sought to warn readers against overzealous CD ROM True Believers.

I claim no intimate familiarity with educational technology or educational technologists. Listening over the professional back fences, however, it strikes me that ET, too, often falls victim to its own True Believers.

Is my perception accurate? And, to the extent that it is accurate, what could and should be done about it? □

Part Six

Bibliography

Interactive Video: Fifty-One Places to Start — An Annotated Bibliography

Doris R. Brodeur

Most of the references listed here have been retrieved from two data bases: ERIC and Microcomputer Index. Each index uses the descriptor "interactive video." Other articles have been retrieved by reading the monthly issues of a few key journals that focus on instructional and video technologies.

Interactive video is a new and rapidly-changing technology. While accounts of early developments are useful and interesting, it is important to keep up with current information. The articles listed here, all published since 1982, represent the major areas of published ideas on the subject of interactive video: hardware and software specifications; features, advantages, and limitations of interactive video; development projects; applications in education, business, medicine, and the military; and experimental, survey, and naturalistic research.

Allen, D. (1985). Linking computers to videodisc players, *Videography*, January 1985, 26-28, 30-32, 34.

Describes the levels of interactivity of videodisc players from the perspective of the hardware manufacturers. Explains that companies and institutions providing products in interactive video generally divide into two camps: those that provide hardware to interface between the computer and the videodisc player, and those that provide both the hardware and software in an integrated system. Includes a list of 23 companies with extensive descriptions of the products (features, models, price, applications) and services that each offers.

Blizek, J. (1982). The First National Kidisc—TV becomes a plaything, *Creative Computing*, 8(1), 106, 108, 110.

Describes the process of designing and producing one of the first interactive discs, "The First National Kidisc." Explains the design requirements that the disc utilize all of the optical videodisc functions: step-frame, still mode, real-time demonstrations, single-frame animation, slow motion, information compression, freeze-frame, and dual audio channels. Provides ten hints for producing videodiscs using optical videodisc players. Includes a listing of the 26 segments of the "Kidisc."

Doris R. Brodeur is Associate Professor, Curriculum and Instruction, Illinois State University, Normal, Illinois.

Boen, L. (1983). Teaching with an interactive video-computer system, *Educational Technology*, 23(3), 42-43.

Reports an experimental design study in which computer-directed instruction (interactive videotape) was compared to traditional lecture for a group of education students enrolled in a course in study skills. Gives eight advantages of using CDI for education and training. Concludes that CDI holds promise as an alternative to traditional classroom instruction.

Bosco, J.J. (1984). Interactive video: Educational tool or toy? *Educational Technology*, 24(4), 13-19.

Presents the arguments of proponents and critics of interactive video on issues of cost, replicability of instruction, responsiveness to human variability, active learner participation, the kind of tasks that require visual information, records of learners' progress, the use of video as a chalkboard, and information storage. Describes the current status of interactive video in consumer and industrial markets as well as in education. Explains the difficulties involved in applying interactive video in educational settings. Emphasizes the need to examine the quality of instruction, not only the potential of the technology.

Broderick, R. (1982). Interactive video: Why trainers are tuning in, *Training*, 19(11), 46-49, 52.

Gives several examples of the application of interactive video to training programs in diverse fields. Describes Hon's cardiopulmonary resuscitation system; the use by Ford Motor Company to train dealers in sales, service, and product knowledge; Los Alamos National Laboratory's program to train technicians in proper protocol for working with plutonium; the U.S. Army's simulations to monitor satellites; and a pharamaceutical company's experimentation for technical training and management development.

Clark, D.J. (1984). Exploring videodisc's potential, *Biomedical Communications*, 12(2), 34-36.

Describes the advantages of videodisc for teaching and data storage in university level biology instruction. Comments on the use of videodisc technology for diagnosis by physical appearance by medical professionals, and for training personnel in many areas of the medical profession. Describes the work of the Library of Congress in transferring archival collections of photographs, animal and plant images, and art collections to videodisc. Explains the advantages and limitations of videodisc technology, and gives examples of generic discs available from a variety of sources. Comments that high cost and hardware configuration problems are the obstacles to rapid adoption of videodisc technology.

Clark, D.J. (1984). How do interactive videodiscs rate against other media? *Instructional Innovator*, 29(6), 12-16.

Provides an introduction to the components of a videodisc learning system: videodisc player, monitor, video-

Interactive Video

185

disc, microcomputer, and computer program. Explains the features of interactive videodisc that will enhance past and current systems of self-paced and computer-assisted instruction. Compares interactive video with lecture, film, slide-tape, videotape, textbook, and conversation, in terms of rapid random access, computer control, branching capability, and learner control. Lists the major obstacles as cost and acceptance by instructors. Looks ahead to future applications of videodisc technology.

Cohen, V.B. (1984). Interactive features in the design of videodisc materials, *Educational Technology, 24*(1), 16-20.

Defines an interactive program as one in which the student is actively involved in responding to the lesson so that some sort of qualitative response is required in order for the instruction to continue. Describes the interactive features which can increase the quality of instruction in computer-based materials, with particular emphasis on interactive videodisc projects: non-linear format of content, user control options, feedback and remediation. Discusses design strategies that accommodate different learner needs: modular organization of content, branching, pretests, case studies which utilize realistic video sequences, and the use of dual audio track. Suggests the use of a video data bank, i.e., skill-oriented still frames coordinated to each module. Concludes that interactivity should be measured not only by the quantity of responses made, but by the quality and variety of responses.

Copeland, P. (1983). An interactive video system for education and training, *British Journal of Educational Technology, 14*(1), 59-65.

Describes the CAVIS interactive video teaching system developed at the West Sussex Institute of Higher Education. Lists nine key requirements of the system that surfaced in the needs analysis. Focuses on two features in the design of the new system: the need to establish a presentation mix and the need for interaction. Describes the hardware and operation of CAVIS that mixes videocassette pictures, text and videotext diagrams, and sound. Explains how the system monitors student performance and system efficiency.

Daynes, R. (1982). Experimenting with videodiscs, *Instructional Innovator, 27*(2), 24-25.

Describes seven sets of videodisc frames used by the Nebraska Videodisc Group in the design of videodisc programming: orientation, content, decision, strategy or comment, summary, problem, and help. Explains the process of producing a pre-mastering videotape in preparation for the final mastering on videodisc. Suggests that videodiscs are different from film and linear video in pace, organization, and style.

Daynes, R. (1982). The videodisc interfacing primer, *Byte, 7*(6), 48-59.

Describes the technical configurations of the levels of

interactivity of videodisc players giving examples of models for each level. Compares, in chart form, videodisc system capabilities for four models of videodisc players. Describes the early work of the Nebraska Videodisc Design/Production Group. Includes illustrations of equipment, diagrams of the interactive videodisc design and production process, and a glossary of 48 terms related to interactive video design and production.

Dennis, V.E. (1984). High tech training at Arthur Anderson and Co., *Instructional Innovator, 29*(3), 9-12.

Describes the training needs at Arthur Anderson Co. that led to the investigation of interactive video as a delivery system. Explains the Technology Life Cycle Model followed by that company in the adoption of any new technology. Includes the time frame for each of the six phases in the model.

Ebner, D.G. *et al.* (1984). Current issues in interactive videodisc and computer-based instruction, *Instructional Innovator, 29*(3), 24-29.

Presents the design strategies used by the Nebraska Videodisc Group. Lists guidelines for determining appropriate graphics and video effects. Addresses such issues as the role of the instructor, the role of the student, time savers, and current research applications. Predicts that future issues will be the role of the instructor, the assessment of students, and the need for experimentation in novel domains.

Ebner, D.G., Manning, D.T., Brooks, F.R., Mahoney, J.V., Lippert, H.T., and Balson, P.M. (1984). Videodiscs can improve instructional efficiency, *Instructional Innovator, 29*(6), 26-28.

Lists six criteria established by a design team to determine what subject matter would be appropriate for an experimental interactive videodisc lesson. Describes the field test of an interactive program to teach intramuscular injections to health sciences students. Explains how the lesson was designed to be used in a teacher-active (as opposed to a student-active) mode. Results showed that students using videodisc learned faster, and retained information longer. Moreover, instructors and students were favorably disposed to the use of the technology.

Fedale, S.V. (1982). Interactive video in the Pacific Northwest, *T.H.E. Journal, 10*(1), 124-126.

Reports the results of a survey to determine the level of interactive video production in the Pacific Northwest. Emphasizes the types of equipment in use, how the equipment is being used within the organization, and opinions of users as to the effectiveness of this system of training and education. Reports that 12 of the 49 respondents are actively involved in the production of interactive video. Describes participant reactions to the questions, rather than statistical data. Concludes that interactive video is being produced in educational institutions as well as in industry.

Fort, W. (1984). A primer on interactive video, *AV Video*, 6(10), 39-42.

Provides an entry-level overview of interactive video. Compares the ease of locating information on videodisc with videotape. Includes diagrams of an optical laser videodisc and a sample flow chart of a branching decision.

Glenn, A.D. (1984). Teaching Economics: Research findings from a microcomputer/videodisc project, *Educational Technology*, 24(3), 30-32.

Describes the development and implementation of an introductory high school economics course funded by the Minnesota Educational Computing Consortium and the Rockefeller Family Fund. Explains that the purpose was to demonstrate that low-cost personal computers and home videodisc players could be used to deliver instruction to students. Reports the results of research on student use of the first unit of the course. Measures indicated that students learned the concepts within the time frame estimated by the developers. Findings indicate that students responded in a positive manner to the experience, and that teachers believed that the materials can be used effectively in a variety of circumstances and classroom settings.

Glenn, A.D. (1983). Videodiscs and the social studies classroom, *Social Education*, 47(5), 328-330.

Explains the features and advantages of videodisc. Describes the applications of videodisc to the classroom: teaching tool, tutor, educational simulation, and storage device/data bank. Concludes that videodisc will be another tool for the teacher to present information, provide remediation, and prepare students for an information society.

Grabowski, B. and Aggen, W. (1984). Computers for interactive learning, *Instructional Innovator*, 29(2), 27-30.

Analyzes the features of computer-based interactive video (CBIV) and matches these characteristics with pedagogical styles of the learners. Examines the internal cognitive processes of the learner (sensory memory, short-term memory, long-term memory) and how they relate to the use of the computer. Suggests that artificial intelligence is the key to a closer match between the technology and learner cognitive processing.

Hannafin, M.J. (1984). Options for authoring instructional interactive video, *Journal of Computer-Based Instruction*, 11(3), 98-100.

Discusses hardware and software alternatives for authoring interactive video. Compares three major alternatives for authoring interactive lessons: authoring systems, authoring languages, and general purpose programming languages. Explains the difference between authoring systems and authoring languages, and lists the advantages and limitations of each alternative. Uses, as examples, BCD's authoring system "The Instructor"; authoring languages, Super-

PILOT, PASS, ADAPT, and McGraw-Hill's IAS; and programming languages, BASIC, EnBASIC, and Higher Text II. Concludes that there are many issues about the design of interactive video that are currently undecided.

Heines, J., Levine, R., and Robinson, J. (1983). Tomorrow's classroom—the changing focus in computer education, *T.H.E. Journal*, 10(5), 100-104.

Traces the evolution of corporate educational services of Digital Equipment Corporation, highlighting its use of computer-based interactive video. Describes the advantages of interactive videodisc that make it a viable approach to individualized personnel training: decentralized training, scarcity of program assistants and instructors, automating adaptability. Explains how DEC uses linear video, computer-assisted instruction, and interactive video in its training programs. Advocates the further development of course authoring systems that will be able to judge the quality of student responses.

Hon, D. (1982). Interactive training in cardiopulmonary resuscitation, *Byte*, 7(6), 108, 110, 112, 114, 116, 118, 120, 122, 124, 130, 132, 134, 136, 138.

Defines the characteristics of the kind of system that was sought to address the needs of cardiopulmonary resuscitation (CPR) trainees. Describes the development of the hardware, software, program design, and peripherals used in the interactive videodisc system. Includes photographs of equipment, video screens, and mannequins, diagrams of performance tests, sample script, and sample programming segments.

Howe, S. (1984). Interactive video, *Instructor*, 93(5), 108-110.

Advocates the use of existing video equipment and microcomputers in school districts to begin to develop interactive programs. Describes the equipment and software needed to plan and write lessons. Explains four types of lessons: insertion of text onto an existing videotape, files of data and visual images that can be accessed at will, simulation, and programmed learning format. Gives an example of a lesson on the history of Philadelphia for junior high students. Includes a list of eight interface device manufacturers.

Howe, S. (1985). Interactive video: salt and pepper technology, *Media and Methods*, January 1985, 8-14, 20.

Explains the three levels of interactivity of interactive video systems, the equipment required to produce programs using videotape-based systems, and the components of videodisc systems. Describes design strategies for four types of programs: file, programmed learning, insertion, and simulation. Provides examples of the current range of interactive video programs. Includes a resource guide of associations and consultants, books, hardware, authoring languages, optical videodiscs, publications, interactive videotape programs, workshops, and a source list.

Jonassen, D.H. (1984). The generic disc: Realizing the potential of adaptive, interactive videodiscs, *Educational Technology*, 24(1), 21-24.

Presents both sides of the conflict between instructional designers and educational publishers over the design of interactive videodisc. Describes the problems of designers: principles of instruction based on aptitude-treatment interaction do not transfer to local applications, and development costs are high. Explains that publishers are limited by the lack of series of programs, and the difficulty of marketing products with a wide range of applicability. Proposess as a solution the generic disc, i.e., instructional video programming and series of still frames organized around a particular subject matter, with the adaptive, instructional designs left to the purchaser. Suggests strategies for local designers: learner control, remediation, grade level/curricular adaptation, prior knowledge, learner styles, and program control. Lists four manufacturers of videodisc authoring systems.

Kearsley, G.P., and Frost, J. (1985). Design factors for successful videodisc-based instruction, *Educational Technology*, 25(3), 7-13.

Describes nine examples of videodiscs to illustrate the range and variety of instructional applications for medicine, the military, higher education, and business and industry. Summarizes the results of a number of evaluation studies that provide evidence that interactive videodisc-based instruction is effective and efficient. Discusses the factors associated with the design of effective videodisc instruction: interactivity, visual design, organization, active participation, metaphors and models, personality, and team approach. Concludes with a list of nine additional design principles that will improve the teaching and learning process.

Keener, J.R. and Bright, L.K. (1983). Improving access to health education, *Health Education*, 14(6), 47-50.

Describes the need for user friendly interactive video systems for family, nursing home, library, and "waiting area" use in order to deliver health information to the general public. Describes the use of interactive videodisc in health education with examples of Hon's CPR program, simulations for medical emergencies, and learning stations for regional nursing homes. Lists the obstacles to the adoption of interactive videodisc. Stresses the importance of faculty development in new delivery systems. Provides examples from the College of Education of the University of Minnesota, Duluth.

Kehrberg, K.T. and Pollack, R.A. (1982). Videodiscs in the classroom: An interactive economics course, *Creative Computing*, 8(1), 98-102.

Describes a development project of the Minnesota Educational Computing Consortium to provide a high school economics course to be delivered by microcomputer and videodisc. Explains that the project was an answer to the problem of declining enrollments, reduction in funds, and scarcity of teacher resources. Provides details on the development and production of the materials. Includes a discussion of the pre-mastering and mastering processes as well as production costs.

Laurillard, D.M. (1984). Interactive video and the control of learning, *Educational Technology*, 24(6), 7-15.

Investigates three basic questions with regard to the balance of student control versus program control in interactive video: the sequence of the presentation, the choice of number and level of practice exercises to do, and the strategy for alternating between the two. Reports using an already existing video lesson that was known to be effective. Describes the structure and content of the interactive lesson and the field trials with 22 students. Concludes that students can make full use of most aspects of control, and designers must demonstrate that they know best what the student needs at each stage. Comments that it would be a misuse of the medium if interactive video were to become as highly controlled and directive as most computer-assisted instruction has been. Includes diagrams of the lesson design and student control options.

Lee, B. (1984). Interactive authoring languages, *AV Video*, October, 1984, 22, 24-25.

Reviews the features, advantages, and limitations of four new interactive video authoring languages. Describes the equipment for which the language is designed to be used. Includes addresses of manufacturers for further information.

Lee, B. (1984). Interactive video in the microcomputer, part II, *AV Video*, May, 1984, 22-25.

Provides a beginning list of contacts, tools, and suppliers of interactive video systems and software. Describes equipment available from Apple, Sony, and AVL, authoring software in the PILOT family, two manufacturers of complete systems, four manufacturers of interface devices, a company that will replicate videotape onto videodisc, and two companies that will pre-master a videotape.

Levenson, P.M. (1983). Interactive video: A new dimension in health education, *Health Education*, 14(6), 36-38.

Focuses on the potential inherent in interactive video and gives examples of present and future applications in health education and health-related fields. Compares the strengths and limitations of current technologies of instruction: the book, video, the computer. Describes features and advantages of videodisc and videotape. Gives examples of programs to teach cardiopulmonary resuscitation, university biology courses, simulations of medical emergencies, identification of muscular disabilities in children, and the use of smokeless tobacco.

Levin, W. (1983). Interactive video: the state-of-the-art teaching machine, *Computing Teacher*, 11(2), 11-17.

Presents an entry level overview of recent developments

in interactive video. Includes information on hardware requirements, and the federal Videodisc/Microcomputer Project to promote courseware development and implementation in the classroom. Explains the advantages of videodisc over videotape, the differences in videodisc formats, and the levels of interactivity. Lists examples of courseware in industry, health, and entertainment. Describes seven development projects with a focus on the classroom. Suggests actions, contacts, and readings for teachers.

Lindsey, J. (1984). The challenge of designing for interactive video, *Instructional Innovator, 29*(6), 17-19.

Suggests that because of the video in interactive systems, there is an affective component to instruction. Recommends heuristics, holistic thinking, persuasion, dramatic tension, and other design techniques that go beyond algorithms.

Lovece, F. (1984). Electronic Learning's April buyer's guide: Videodisc hardware, *Electronic Learning, 3*(7), 60-65.

Presents a summary of the features offered by currently available video interface devices and laser disc players, and a brief description of how they work. Illustrates in a diagram how a videodisc player and an interface device are hooked up to other components in an interactive video mode. Includes a chart of eleven manufacturers of interface devices comparing them for price, compatible computers, compatible video equipment, authoring software, and documentation. Concludes with a chart of four manufacturers of videodisc players, comparing them for price, compatible interface devices, and customer support.

Manning, D.T., Ebner, D., Brooks, F.R., and Balson, P. (1983). Interactive videodiscs: a review of the field, *Viewpoints in Teaching and Learning, 59*(2), 28-40.

Analyzes the advantages and present problems of videodisc format, production, and use. Discusses the major decision or task areas in authoring and producing videodiscs: selection of media and the delivery system, instructional strategies, mock-up or simulation, pre-mastering, mastering, authoring team, and evaluation. Gives examples of current uses in education and training as well as an evaluation of specific applications. Predicts future trends in technical developments, in technical programs, and in military and industrial training. Includes a list of 84 references.

Martorella, P.H. (1983). Interactive video systems in the classroom, *Social Education, 74*(5), 325-327.

Explains how an interactive video system combines the power of the microcomputer to manipulate variables and retrieve information, with the capability of video to present varied visual materials. Describes three Social Studies projects employing interactive video systems: Project CENT, a consumer economics course; a MECC Project, an economics course for secondary level; and Project ComICon—Computers to Individualize Concept Learning,

supported by Atari, to develop prototype materials to encourage teachers to develop their own interactive materials.

Meyer, R. (1984). Borrow this new military technology, and help win the war for kids' minds, *American School Board Journal 71*(6), 23-28.

Explains that the military uses interactive video for training because it is portable, standardized, makes effective simulations possible, and is a feasible way to provide hands-on experience to trainees. Describes programs in the military curriculum designed to teach leadership, counseling, and personal problem solving. Gives an example of how a program of the Spatial Data Management System is being used in a public school system to teach students how to gather information, study, and take tests.

Parker, W. (1984). Interactive video: Calling the shots, *PC World*, October, 1984, 99-108.

Lists authoring systems, authoring languages, and interface devices available for use with the IBM-PC. Describes projects developed by users in a variety of applications. Compares videotape and videodisc. Explains and gives examples of the features of authoring systems: presentation of information, testing, applications simulations, record keeping. Includes a flowchart of an interactive video program segment.

Price, B.J. and Marsh, G.E. (1983). Interactive video instruction and the dreaded change in education, *T.H.E. Journal, 10*(7), 112-117.

Advocates the use of computer and television technologies as a solution to some of the economic problems faced by public schools. Describes the development of courses in trigonometry, chemistry, and physics using computer-controlled videotape. Emphasizes that these are complete courses for use without the assistance of a teacher. Includes a diagram of the design process used in mapping each course: content, instructional design, and video production.

Reinhold, F. (1984). How they're using interactive videodiscs, *Electronic Learning, 3*(7), 56-57.

Reports the applications of interactive video in six schools that participated in the Videodisc-Microcomputer Project, sponsored by the American Institutes for Research and the U.S. Department of Education. Projects include: a small parochial school in Illinois which developed a program to teach time signature to music students; a district media center in Minnesota which developed lessons for elementary students in economics, punctuation, dictionary uses, and animal classification; a school for the deaf in California which developed programs in language arts and reading; a teaching laboratory school in Florida that developed interactive programs using the videodisc "Whales"; and two universities in Pennsylvania and Nebraska with projects in percentage measurement, chemistry, and biology.

St. Lawrence, J. (1984). The interactive videodisc here at last, *Electronic Learning, 3*(7), 48-54.

Gives a background on what makes a videodisc interactive and goes on to list ten steps in preparing a videodisc. Includes a chart of 33 educational videodiscs, listing title, curriculum area, grade level, format and number of discs, price, running time, description, vendor, and comments. Discs cover subjects in elementary and secondary computer literacy, art, photography, music, English, Spanish, history, economics, science, math, career guidance, and health and recreation.

Schlieve, P.L. and Young, J.I. (1983). How to program interactive learning programs, *Instructional Innovator, 28*(2), 28-33.

Describes two main types of development strategies in the design of interactive video (both videotape and videodisc): subject matter organization and learner involvement. Explains the design process: instructional objectives, performance assessment, program design, and script preparation. Presents five basic flowchart patterns of learner involvement in a lesson. Highlights the implications of interactive video for education.

Simcoe, D. (1983). Interactive video today, *Instructional Innovator, 28*(8), 12-13.

Surveys a random sample of 100 media center directors and administrators of universities, colleges, and technical schools in New York State for their attitudes and opinions of interactive video instructional systems (IVIS) in higher education. Reports that attitudes are positive, with over 25 percent already using some form of IVIS. Mentions that software development and financing are seen as the major barriers. Recommends that media directors keep current with software development techniques.

Thorkildsen, R., Allard, K., and Reid, B. (1983). The Interactive Videodisc for Special Education Project: Providing CAI for the mentally retarded, *The Computing Teacher, 10*(8), 73-76.

Describes three major projects conducted under the auspices of the IVSET program: the development of the hardware and software components required to produce CAI materials for the mentally handicapped, a bilingual computer-aided assessment instrument for mathematics skills, and an instructional program to teach social skills to behaviorally handicapped students. Discusses the findings of field tests on the issues of instructional development, mentally handicapped populations, system reliability, and cost-effectiveness. Concludes that interactive videodisc systems have great potential for assessment, diagnosis, prescription, and instruction of handicapped and normal students.

Thorkildsen, R. and Friedman, S. (1984). Videodisc in the classroom, *T.H.E. Journal, 11*(4), 90-95.

Describes videodisc technology, its advantages over similar media, levels of interactivity, and comparisons of videotape and videodisc. Provides examples of its use in industrial, medical, and military training programs. Describes development projects and materials for educational applications: the circulatory system, "Schooldisc," a MECC economics course, a university art history project, introductory physics, legal education, and space science.

Thorkildsen, R., and Hofmeister, A. (1984). Interactive video authoring of instruction for the mentally handicapped, *Exceptional Education Quarterly, 4*(4), 57-73.

Describes the development and field testing of a project for the mentally handicapped, sponsored by the U.S. Department of Education and the Interactive Videodisc for Special Education Technology (IVSET) project of Utah State University. Emphasizes the development of the authoring system used to create the interactive materials. Reviews the limitations of CAI research results for the handicapped population.

Utz, P. (1984). The early stages of producing an interactive videodisc, *AV Video, 6*(11), 40-43.

Explains the advantages and limitations of the three levels of interactivity of videodiscs and the features of the various level videodisc players. Describes a five-step design process: assessment, audience, media selection, flow-chart design, and script and storyboard. Includes a sample flow-chart design for a nursing program videodisc.

Utz, P. (1984). So you want to make a videodisc?, *AV Video, 6*(10), 34, 36-38.

Explains questions to ask when contemplating the production of an interactive video program: interactivity, random access, cost, number of copies, and program revision. Describes the equipment required and gives an example of equipment specifications with prices for a Sony interactive videodisc package. Includes a glossary of 22 common videodisc terms.

Young, J.I., and Schlieve, P.L. (1984). Videodisc simulation: training for the future, *Educational Technology, 24*(4), 41-42.

Describes two projects that use videodisc simulation for training: communication operators in the Signal Corps who were taught to operate complex pieces of equipment and cardiopulmonary resuscitation taught with videodisc simulation and a mannequin. Lists 15 questions to determine if videodisc simulation is a viable alternative to an instructional problem.

Zollman, D. (1984). Videodisc-computer interfaces, *Educational Technology, 24*(1), 25-27.

Lists and describes interface devices that are available for the design and production of interactive video, using videodisc technology. □

Article Citations

1. Interactive Video: Educational Tool or Toy? By James J. Bosco. *Educational Technology*, April, 1984, pages 13-19.

2. Adopting Interactive Videodisc Technology for Education. By Peter Hosie. *Educational Technology*, July, 1987, pages 5-10.

3. Interactive Lesson Designs: A Taxonomy. By David H. Jonassen. *Educational Technology*, June, 1985, pages 7-17.

4. Learning the ROPES of Instructional Design: Guidelines for Emerging Interactive Technologies. By Simon Hooper and Michael J. Hannafin. *Educational Technology*, July 1988, pages 14-18.

5. Improving the Meaningfulness of Interactive Dialogue in Computer Software. By George R. Mc-Meen and Shane Templeton. *Educational Technology*, May, 1985, pages 36-39.

6. Designing Interactive, Responsive Instruction: A Set of Procedures. By Maria Harper-Marinick and Vernon S. Gerlach. *Educational Technology*, November, 1986, pages 36-38.

7. Interactivity in Microcomputer-Based Instruction: Its Essential Elements and How It Can Be Enhanced. By Herman G. Weller. *Educational Technology*, February, 1988, pages 23-27.

8. Design and Production of Interactive Videodisc Programming. By Eric A. Davidove. *Educational Technology*, August 1986, pages 7-14.

9. Project Management Guidelines to Instructional Videodisc Production. By Martha R. Tarrant, Luke E. Kelly, and Jeff Walkley. *Educational Technology*, January, 1988, pages 7-18.

10. Designing a Visual Factors-Based Screen Display Interface: The New Role of the Graphic Technologist. By Tony Faiola and Michael L. DeBloois. *Educational Technology*, August 1988, pages 12-21.

11. Design Factors for Successful Videodisc-Based Instruction. By Greg Kearsley and Jana Frost. *Educational Technology*, March, 1985, pages 7-13.

12. Educational Strategies for Interactive Videodisc Design. By David Deshler and Geraldine Gay. *Educational Technology*, December, 1986, pages 12-17.

13. Using Interactive Video for Group Instruction. By William D. Milheim and Alan D. Evans. *Educational Technology*, June 1987, 35-37.

14. A Videotape Template for Pretesting the Design of an Interactive Video Program. By Scott V. Fedale. *Educational Technology*, August, 1985, pages 30-31.

15. Visuals for Interactive Video: Images for a New Technology (with Some Guidelines.) By Roberts A. Braden. *Educational Technology*, May, 1986, pages 18-23.

16. Storyboarding for Interactive Videodisc Courseware. By James F. Johnson, Kristine L. Widerquist, Joanne Birdsell, and Albert E. Miller. *Educational Technology*, December, 1985, pages 29-35.

17. Videodisc Simulation: Training for the Future. By Jon I. Young and Paul L. Schlieve. *Educational Technology*, April, 1984, pages 41-42.

18. Design Considerations for Interactive Videodisc Simulations: One Case Study. By Stephen R. Rodriguez. *Educational Technology*, August, 1988, pages 26-28.

19. A Model for the Design of a Videodisc. By Joanne C. Strohmer. *Educational Technology*, April, 1988, pages 38-40.

20. Interactive Video and the Control of Learning. By Diana M. Laurillard. *Educational Technology*, June, 1984, pages 7-15.

21. An Analysis of Evaluations of Interactive Video. By James Bosco. *Educational Technology*, May, 1986, pages 7-17.

22. A Comparison of the Effects of Interactive Laser Disc and Classroom Video Tape for Safety Instruction of General Motors Workers. By James Bosco and Jerry Wagner. *Educational Technology*, June, 1988, pages 15-22.

23. How Effective Is Interactive Video in Improving Performance and Attitude? By David W. Dalton. *Educational Technology*, January, 1986, pages 27-29.

24. The Validation of an Interactive Videodisc as an Alternative to Traditional Teaching Techniques: Auscultation of the Heart. By Charles E. Branch, Bruce R. Ledford, B.T. Robertson, and Lloyd Robison. *Educational Technology*, March, 1987, pages 16-22.

25. Interactive Video: A Formative Evaluation. By Roberts A. Braden. *Educational Technology*, September, 1985, pages 33-34; October, 1985, pages 32-33.

26. A Videodisc Approach to Instructional Productivity. By Larry Peterson, Alan M. Hofmeister, and Margaret Lubke. *Educational Technology*, February, 1988, pages 16-22.

27. History in the Making: A Report from Microsoft's First International Conference on CD ROM. By David Rosen. *Educational Technology*, July, 1986, pages 16-19.

28. Anticipating Compact Disc-Interactive (CD-I): Ten Guidelines for Prospective Authors. By Michael DeBloois. *Educational Technology*, April, 1987, pages 25-27.

29. CD ROM Joins the New Media Homesteaders. By David C. Miller. *Educational Technology*, March, 1987, pages 33-35.

30. Interactive Video: Fifty-One Places to Start—An Annotated Bibliography. By Doris R. Brodeur. *Educational Technology*, May, 1985, pages 42-47.

Index